T0256974

A GENIUS PLANET

APPLIED VIRTUALITY BOOK SERIES VOL. 11

A GENIUS PLANET — ENERGY: FROM SCARCITY TO ABUNDANCE, A RADICAL PATHWAY

LUDGER HOVESTADT,
VERA BÜHLMANN,
SEBASTIAN MICHAEL

BIRKHÄUSER
Basel

SERIES EDITORS

PROF. DR. LUDGER HOVESTADT
Chair for Computer Aided Architectural Design (CAAD),
Institute for Technology in Architecture (ITA), Swiss Federal Institute of Technology (ETH),
Zurich, Switzerland

PROF. DR. VERA BÜHLMANN
Chair for Architecture Theory and Philosophy of Technics, Institute for Architectural Sciences,
Technical University (TU) Vienna, Austria

LAYOUT AND COVER DESIGN: onlab, CH-Geneva, www.onlab.ch
TYPEFACE: Korpus, binnenland (www.binnenland.ch)
PRINTING AND BINDING: Kösel GmbH & Co. KG, D-Altusried-Krugzell

Library of Congress Cataloging-in-Publication data
A CIP catalog record for this book has been applied for at the Library of Congress.

Bibliographic information published by the German National Library
The German National Library lists this publication in the Deutsche Nationalbibliografie; detailed
bibliographic data are available on the Internet at http://dnb.dnb.de.

This publication is also available as an e-book
ISBN PDF 978-3-0356-1421-3
ISBN EPUB ISBN 978-3-0356-1419-0

© 2017 Birkhäuser Verlag GmbH, Basel
P.O. Box 44, 4009 Basel, Switzerland
Part of Walter de Gruyter GmbH, Berlin/Boston

Printed on acid-free paper produced from chlorine-free pulp. TCF ∞
Printed in Germany

ISSN 2196-3118
ISBN 978-3-0356-1406-0

9 8 7 6 5 4 3 2 1

www.birkhauser.com

TABLE OF CONTENTS

FOREWORD

This book has a simple and optimistic message: energy isn't a resource, energy is clean, and energy isn't scarce, in fact the opposite, it is abundant! Because now, with information technology, energy has become what we might call an 'intellectual wealth' that can be captured, stored, distributed – 'cycled', so to speak – by electronic coding. And as there are no limits in principle to how much 'energy cycling' is possible, energy itself loses the limitations we're used to associate with it.

It is not, at first, entirely obvious or perhaps intuitive to think of energy as an 'intellectual wealth' that can be 'cycled', and so the book explains in detail how and why this is so, and it makes a compelling case for embracing this extremely relevant reality: we have more than enough energy. For the foreseeable future, and beyond. We can relax.

The book has a complicated story. We started writing it 8 years ago, at the beginning of 2009, and completed it in 2011. We are researchers at a renowned university.[1] Our joint background covers architecture, information technology and philosophy. This was our first book collaboration, and in a sense it formed the basis for a whole book series that has since appeared. Our overriding thesis is that through information technology, and, more specifically, through the mathematical notion of information that makes information technology possible, we find ourselves in a new world.

The expression 'a new world' may seem to have a whiff of a 'New Age' sentimentality about it, but that's absolutely not what we mean. Every major step in our development has been brought about by a categorically new understanding, which therefore enabled radically new techniques and, by extension, technologies. Take, for example, the beginning of modernity, around 1400 CE. By this time we'd learnt how to symbolise zero and had integrated it into the corpus of our mathematical thinking, thus laying the foundations for the Renaissance.

'Mathesis', from which we get mathematics, in Greek simply means 'that which can be learnt'. It is in this way that we think we should begin to address the world we live in as a new world: a world in which things that before were simply out of the human scope can be learnt and mastered intellectually.

Much as at those previous junctures, we recognise today that old, established orders are being turned upside down. Possibilities are boundless, trusted certainties disappear. That's why, across the globe, we find ourselves afraid and justifiably cautious. But fear is a bad advisor; it is an amplifier and accelerator, it magnifies crises, whether they are perceived or real, and in doing so helps bring them about faster, and carrying with them more devastation, than otherwise they might. Caution, meanwhile, also demands courage, because without courage it turns into timidity and inaction. And being timid and inactive is not the same as being relaxed.

With this first book, *A Genius Planet*, we emancipated ourselves from the reductionism — more fearful than cautious — of the prevailing discourse of crisis: sustainability, global warming, world famine, The Empire... Together with author and playwright Sebastian Michael we wrote this book in an accessible language and believed this positive message of an emancipation from the constraints of imagined limitations would be welcome. We quickly had to learn that the opposite was the case: we found ourselves the target of massive hostility, and, even after many repeated attempts by professional and well connected literary agents, were unable to find a publisher. This is all the more astonishing in view of the fact that the things we discuss in the book are now one by one materialising, quite in contrast to the dystopias that have been, and continue to be, conjured up elsewhere.

Today, and with the wisdom of hindsight, we may call our intentions with this book naive: to spell out, in direct language, that energy is all around us, principally in abundance, and that it is clean; that it is a mistake to think of energy as coupled with resources and scarcities. It was naive to think that some simple facts, which can easily be verified, together with an outline of the principal technologies, which are readily available, well put together could make a 'popular' book; that an unmediated optimistic explanation could get a hearing. That it would be considered acceptable — permissible, even — to say: 'Look, it's really simple. Why create so many problems on the basis of assumptions that were correct 200 years ago, but that today are so obviously inapplicable?'

Where did we go wrong? As researchers we had underestimated how radically information technology has turned our world upside down. Today it seems impossible to write in order to enlighten. Writing today is used principally to synthesise our world. It isn't about messages any more, what's in demand are prophecies. Research, theory and technology no longer occupy a secure, neutral space: they are no longer capable of serving as a reference for certainty. Instead, research, theory and technology find themselves on a wide open, unknown digital plane. It is no surprise that people are scared of this boundless open space, from which a cold wind seems to blow: you have to advance slowly and cautiously. You have to find consensus, establish trust and enter compacts — in other words, do politics — to protect yourself against the unknown, as far as this is even possible, step by step. Nobody, not scientists, nor technology itself, can tell what is right or where the journey might lead. So we cling to the familiar, the old, that which we've brought along from the past. We hold on to a formal logic, insisting that 'the calculation of probability' is an illusion, used only by those whose aim it is to conceal true reason; we cling on to constants in nature, to territory, to our supposed 'natural' roots. We huddle together in small groups and nod at each other, agreeing with each other that we have to stick together and fend off the 'evil' that threatens from the outside. We conspire. Out of fear, we overlook, or even sacrifice, more or less deliberately, the possibilities this new world affords, the avenues that open up, and the methods that offer themselves for us to find our way around it. As a result, our world gets obscured in unquestioned performative ritual: our words and our symbolic practices do not just convey meaning, instead, by being expressed, they create their own 'reality', which we then, because we have just affirmed this reality in expressing it, take at face value. And so, ostensibly pragmatic, we all too often and all too readily sacrifice what is fragile, beautiful, optimistic, just to allay our fears. Why do we get all these dystopian news stories, disaster movies and endgame theories today? That's a question we should have asked ourselves when we set out to write *A Genius Planet*.

Let's take, simply as an example and thought exercise, climate change, and for a moment follow freely and without prejudice the argumentation of someone who says pretty much the opposite of what everybody else is saying. We know what everybody else is saying, so we are interested in this near-lone voice of dissent: physicist and Nobel laureate Ivar Giaever.[II] He asks the question, 'is it even possible still, today, to talk about global warming?' And he quickly realises: 'no, it isn't!' Because if you do, you will immediately be branded an irresponsible climate change denier. Many have noted as he has – and with some degree of frustration – that the climate change discussion has taken on a religious tone. We realise, of course, that this is highly contentious, but you could also think of Global Warming as a modern myth: a narrative, with which people and cultures express their understanding of their world, of the challenges they face, and of themselves. Myth demands that the truth it claims be taken seriously. Which is what we observe in the climate change discussion. Religious myth ties the human existence to the world of the gods. Is doubt in god allowed? No. Is doubt in global warming allowed? No. But is there really – quite in contrast to god – anything left to discuss about global warming? Well, according to Giaever, surprisingly little. He finds that measured from space, global temperature looks remarkably stable: over the last 150 years it has changed by 0.3 %, and in the last 30 years not at all.[III] That, to him, makes the 'global warming debate' sound silly: if these findings are correct – and as scientists we have to at least allow for the possibility that they are – then the overall temperature on planet earth is not going up at an alarming rate, it's hardly going up at all. This stands in the starkest contrast to what almost every other scientist thinks is the case. Yet even if it is true, does this mean climate change isn't happening? Of course it doesn't: we can clearly see that climates do change across the globe: there are obvious shifts taking place. But that's a different matter to the overall temperature increasing, and it therefore calls for a different approach.

So let's, again just for the sake of argument, try one such different approach: say the global temperature is quite stable, but in different parts of the world the climate is undergoing rapid and significant changes. What this really means is that weather patterns change. And they do so all the time. But what do we talk about when we are afraid? We talk about that which scares us. We talk about the bad days with the bad weather, and over an extended period we talk about the bad years. What do we talk about when we're in love? We talk about the things that make us feel good: the good days with the good weather; the good years. The weather, much as our mood, is in constant flux. So how is it that we have statistics in which the temperature doesn't change much (good mood), which nobody wants to read, and at the same time we have statistics in which the temperature increases dramatically (bad mood), which everybody reads? If Giaever's data is accurate, then global temperatures don't change, what changes is regional climate and local weather. But very obviously, if, globally speaking, we find ourselves 'in a fearful mood', we end up talking only about the bad weather, whilst overlooking the good weather that also exists somewhere else on the globe. And these numerous instances of 'bad weather' are what we then mistakenly call 'global climate'.

Yet it gets worse: because we are 'in a bad mood', and because we are scared, we go to extraordinary efforts, setting up new weather stations precisely where the 'bad weather' is. What ensues are self-fulfilling prophecies; we confirm our cause for fear: the Arctic is getting warmer. And that doesn't need to be argued about either: we have set up most of our new weather stations over the last thirty years in the Arctic, and we can see from them that it's really, genuinely, getting warmer there. It's where we focus our digital attention, and we have plenty of statistical as well as anecdotal and pictorial evidence to support and illustrate our worst fears: polar bears floating helplessly on melting slabs of ice; majestic ice shelves breaking up off the Greenland coast; glaciers receding in the Alps. None of this is in any doubt. But what we overlook is the 'good news': the Antarctic is comparatively stable. Yet in the entire southern hemisphere Giaever notes a fraction of the number of weather stations there are in the northern hemisphere. So could it be that we avoid having a closer look, because, paradoxically, the climate is actually quite healthy there? Could it be the case that there is also 'good news', but that this does not fit into our current narrative, but upsets everything? Any certainty we may have had about an impending global disaster gets most inconveniently differentiated by a separate set of data, which suggests: there are shifts and changes, but they are not what we have already agreed they must be. So we sacrifice this 'good news' on the altar of our fearful consensus, which we've established by staring at the ice that melts away from the Arctic. This would mean that it is precisely *because* we've been huddling together for thirty years now, fearfully looking in just one direction and adding so disproportionately to the number of surveillance points there that to our minds the world temperature is rising at an alarming rate. And so we end up with two sets of statistics: measured on the ground, world temperature increases rapidly. Measured from space, it almost stays as it was. That's how Giaever argues, and he wonders what the fuss is about.

The fuss, of course, is about the adjustments we have to make to the big changes that are indeed happening. Perhaps, then, we should not worry about the 'world climate' (there really is no such thing as a 'world climate', there are different climates, constantly changing, all over the world), but about the apocalyptic prophecies that instil fear in us and poison the discussions that are genuinely needed to address these serious 'bad weather' phenomena, which *do* threaten us and which claim the lives of thousands of people, every year.

Does this mean then that we should reject the myths and distortions, banish them as dangerous or wipe them under the carpet as nonsense, demanding instead dispassionate reason and logic? That would do these myths an injustice; and in any case, they are too powerful, too strong. And, that aside, we would then be pandering to yet another myth, the myth of absolute reason, a reason without roots in history or experience, which is something we ourselves might take many a physicist to task for.

No, we cannot reject the myths in themselves. They are the poetry of legends. And we need not look far to find their potency: the speculations of Homer[IV] or Hesiod,[V] for example, they are too masterful to be rejected. Reason or logic might tell us to do so, or to belittle them, but they contribute too much to our culture to be ignored. Instead, we need to learn to handle these myths, to integrate them into our thinking: embrace them, re-imagine them, coat them informationally, allow them to be part of us, much as an immunisation might be a benign dose taken to better equip us against any potentially malign force that could damage, at worst and over time even destroy, us.

There are, of course, many ways of looking at this. But why not spin this thought about mythology a little further, for example by drawing the line from Homer and Hesiod to Xenophanes.[VI] Although he was highly critical of them, it was, for him, not so much about rejecting their myths, but about affirming these myths in their place. He separated knowledge from belief and asserted human intelligence as the source of science and discovery – of knowledge – rather than the gods. This on the one hand emancipates us human beings from the myth, while it at the same time also frees up the gods to play their proper role as symbolic expressions of meaning, rather than as the meaning itself. Thus, you could

A GENIUS PLANET

argue, Xenophanes introduces an ordered mind. And today, similarly, only an ordered mind is capable of conducting a political discourse and channeling the chaotic storm of everything going on at the same time on the digital plane into structures we can make sense of. The ordered mind now is what we may call a *character* who takes on responsibility and a personality. The order, the structured channels, in turn, are what we may call an *economy*.

Within this analogy, the first few people to economise the many things with an ordered mind are called *tyrants*. They do not follow the myth, rather they stand up to it. They take a bold step out of the myth and challenge it, negotiate with it the possibilities, enter contracts with it. In doing so, they become *indebted* to the myth. Thus they economise the myth, use it too, but don't follow it. They promise to 'repay their debt with interest': it's a complicated relationship.

The fearful multitude, by contrast, tend to subject themselves to nature, to the outside forces, the technologies, the powerful, the tyrants, and follow them. They make possible the tyrannical imposition on their person, or rather, they offer the *sacrifice* of their person, so as to be without debt. Instead, they invest and hope for a return. Today we have many myths, many ritual sacrificial altars: terrorism, world famine, national debt, the scarcity of energy. These monsters petrify us. We don't want to become indebted ourselves, and that is exactly what makes it possible for these myths to reach into our thoughts and actions without hindrance. We yield, without meaning to, without questioning, and without control, up our 'credit'. Manipulated from the outside, we turn ill-tempered and aggressive against everything 'other': we become inhospitable, xenophobic, racist. Our personal world gets dull. We lose our ability to see in the other, in the unknown, possibilities, beauty and challenge. Our world becomes mechanical, athletic, pornographic. It loses all charm, all eroticism. We even lose our ability to enjoy fine weather on a fine day. The world could be our guest, but we have no home, no identity: we are nobody. So it's no use if we then go complaining to the tyrants, bring our sacrifice and offer our 'investment', hoping that things will turn out all right for ourselves. Instead, we have to put our mind in order, learn to talk and 'invest' our spirit – our 'capital' – well: give credit, take on debts, enter contracts and square up to the tyrants. That's how politics comes about. That's how we become a character, a person.

What could this mean in concrete terms? Well, drawing the arc back towards the example of climate change, it may mean allowing ourselves to ask a different kind of question altogether. For example, can there be such a thing as an incontrovertible truth in science? Or, what is the optimal temperature for planet earth? Is it necessarily what it is now? That, Giaever thinks, 'would be a miracle'. Or, faced with obvious climate changes, are they unequivocally a bad thing? For some communities, such as those on pacific islands that lie just above a rising sea level, who are about to lose their entire habitable place on earth: absolutely. They are faced with an existential crisis. But are communities that have for centuries been living surrounded by, and with, ice necessarily worse off if that ice now is melting and new fauna and flora comes to inhabit their lands and seas? There is no doubt that behaviours, customs, hunting traditions, diets will have to change. But is that without question by definition a 'bad thing'? Could it not, quite conceivably, offer new opportunities, easier lifestyles or – dare we suggest it – exciting economic openings? Or, what if all this climate change is real, is man-made, is causing radical new realities, but, in the longer term, these new realities are not as catastrophic as we are made to believe, for example because of new technologies? [VII] Or could it be that climate change is real, is man-made, is causing radical new realities, is devastating for many parts of the world, but is simply not our biggest problem, and we would be better off addressing other problems with much greater priority, for example poverty? [VIII]

We don't have the answers to these questions, nor – we want to emphasise – do we take a particular stance or put forward any of these specific views or perspectives. But we do advocate asking questions such as these, and we defend vigorously the right and ethos that not just allow us, but embolden us and each other to square up to tyrants, whomsoever they may be. Many a truth is inconvenient, but none is ever absolute.

Back then to *A Genius Planet* and the question of energy. We know – and that's the subject of our book – that energy is principally available in abundance: it streams from the sun and we can harvest it in untold 'quantities' directly as electricity by means of photovoltaics. This is the good news that we sacrifice on the altar of fear, a fear that stems from our perception of energy as something scarce.

With this quantitative evaluation of energy, we open every cultural and political floodgate around the world, and immerse the world in energy that gets cheaper all the time. This brings about a whole new realisation, which is that our only protection from such a flood of energy – our only way to *handle* this abundance – is a cultural qualification of energy: what should be foremost on our minds is no longer the question 'how much energy do we have', or 'how much does energy cost', but 'what do we want to do with our energy?' Because we will not be able to meet this energy surplus with ideology or economy, not with necessities, but only with *politics*. As we write this, we witness in the Middle East the cruel distortions that come about when established economic necessities disappear (a massive drop in the oil price), and observe how new technologies (fracking, renewable energy) open up a political space that apparently cannot be occupied in a peaceful way because we lack both the political agents and any adequate vocabulary to do so. What we are faced with is a chaotic clash of all kinds of myths and technologies that are not understood, jumbled up with good intentions in an unholy mix that features corrupt raids on resources and religious doctrine. The consequences, unsurprisingly, are horrific.

What then to do? How about we open ourselves to the world's abundance, invite the world to our table. Allow it to school our intellect. So that we no longer have to follow the tyrants economically, but become able to challenge them. That is our ask: no more and no less. Things become unattractive at a mechanic pace, a pace we don't want to keep up with and follow blindly. What we want to do is take things as they are, and let them revolve on their own terms. Everything – nature, culture, art, technology – thus is our guest and bears rich gifts. Richer than we may grasp, more dangerous than we can bear.

And that's why it was a mistake, driven by good intentions, to write a popular science book with an optimistic, enlightening stance. It simply is a contradiction in terms. Today, for all the reasons we've outlined above, and sad though this may be, only threats can be popular. But for us, this book was nevertheless a liberation. Through writing it we have honed our thinking, and we've continued to do so with a whole series of books. Especially the titles *Sheaves – When Things Are Whatever Can Be the Case* and *A Quantum City – Mastering the Generic* are, in a sense, a culinary feast of the intellectual riches of our world.

Now, after a journey lasting eight years, we have decided to publish our first, problematic, book *A Genius Planet* as part of this series. Just as it is and in spite of the fundamental error that we have identified in our stance. We trust you, our readers, to see beyond this. Because *A Genius Planet* still, we believe, opens an important perspective on the energy discussion, and with its unique gesture of fusing pragmatic application, theoretical backdrop and accessible storytelling, we also think it still has a valid contribution to make. And so we hope that this book too, and in spite of its positioning error, will be fruitful and find its way.

What we have done is brought the figures, statistics and references up to date, so that when it appears now in 2017, *A Genius Planet* will still give you a current picture of the world we are talking about.

We are outraged about much of what happens in this world; much we find great joy in, and there is much that we, like everyone, simply avoid. We were born into this world, but we don't feel we're in safe hands; we are in love, but we don't commit. We don't agitate, but we have something to say…

Zürich, Vienna, London, October 2016

GENIUS   I

Imagine a world where there isn't enough energy, but *more than enough.* More than enough food too, and water. More than enough room for everybody who is here now, and for everybody who's going to join us in the foreseeable future.

Imagine, if you will, a world of plenty. Where we have more than we need of everything; a world where we have arguably *too much:* so much perhaps, that we can't be entirely sure what to do with it all; too much at any rate to worry about things like running low on resources or not having enough arable land to feed everyone.

And imagine this being possible without polluting the environment, without overheating the atmosphere, without destroying the basis of our existence. After all: what would be the point of plenty if it killed us? Imagine a world of plenty that could last and prosper.

It would be a different world to the one we're used to. We should find it easy, in such a world, to take care of the rain forests and we would not need to have grave concern for the welfare of polar bears. We would have a different attitude to nature and our environment altogether, because we would not see our own comfort and wellbeing pitched against that of the planet as a whole. And it could be a world in which we would not be governed by fear. We would not have to worry about the population size of our own species and we wouldn't have to make it one of our main concerns to meet our needs with *less.*

Consequently, we wouldn't be ruled by need either: meeting our needs would not be foremost on our minds, we would take our needs being met for granted and concentrate on what we *want* to do, what we *can* do. We would be able, you might say, to *cultivate* our use of energy and of resources, because we wouldn't be thinking in terms of survival, we would be thinking in terms of possibilities. We would, to put it quite simply, be able to *relax.*

That's the kind of world this book is about. A world of plenty, a world, in fact, of more than enough. A world where we can relax and dream up entirely new possibilities, and think, as a result, on a different scale. This is perhaps the most important point we want to make: we can think about our situation differently. Not a little differently, but on a categorically different level.

We know that the first reaction from many of you reading these opening paragraphs will be an instinctive tinge of doubt that will lead you, in the first instance, to ask for figures and facts. For comparative studies. For pie-charts and stacked bars. And we are going to give you some figures, we will illustrate with some scenarios covering opposing ends of the global spectrum that what we are talking about in this book is maybe fantastical, but certainly not fantasy.

But first of all, allow yourself a leap of the imagination. Untie yourself from everything that statistics, forecasts and news reports have told you about energy over the last thirty years. Hardly any of them will be relevant for very much longer, so leave behind, for the time-being, the kinds of figures you are used to, and imagine *not* a world where ten billion people can just about scrape by, but a world where *a hundred billion* people can live comfortably, with education, health care, a balanced diet, central heating, air conditioning, high-speed internet and efficient transport. Because *that's* the kind of gear shift we want to be talking about. Not marginal percentiles, but *orders of magnitude.*

In such a world, we would think really differently, about everything. Not just about energy, water and food, but also about each other, about migration, about the role of nation states, about urban planning, about warfare, property, sharing. About all the things that matter to us and that have agitated us for centuries. About our very *ability* to do something meaningful with our existence, meaningful at least to us and future generations.

Would this be a perfect world, a Nirvana, a Paradise on Earth: Utopia? Of course not. There would be a multitude of new, gigantic challenges waiting for us to deal with, not least some fundamental questions about our ethics and how we organise our societies in a world where 'everything is possible'. Take this to the extreme, if you like, and ask yourself: if there are no limits to what we can do, what *will* we do? These are not trivial matters, they go to the heart and essence of who and what we are as human beings.

So we would certainly have many grave matters to address, but we could do so with a new mindset: our politics, our human interaction, our business practices would all take on a new hue. And would *that* necessarily be a good thing? We think so. We think that it would be a definite step forward. Which is why we have written this book. We don't think that there is an end to all of our problems, with serene bliss and pure happiness waiting just around the corner, but we do think that there is another perspective to be taken on the energy debate. And energy is at the core of every single aspect of human life, which is also why, even when we propose to take a broad, in parts conceptual view with this book, we will still very much focus on the energy question.

And what *is* the energy question?

This is the most important point to get right, right from the start. Because, very obviously, if we keep asking the wrong question then we will invariably keep getting the wrong answer. The energy question is not one of quantity. It is not: 'have we got enough energy/water/room for everyone in the world, and will this continue to be the case if there is more growth?' The question is one of organisation: 'how are we going to organise the plentiful energy/water/room that we have on planet earth, so everyone can live life to the full of their potential?'

There is at our disposal today, with technology that you can buy now and start using immediately, abundant, clean energy. It's almost impossible to overstate just how significant this is. What we are describing in this book is not speculative, we do not have to wait for some promising inventions to come good. Nor do we have to lobby governments or wait for officialdom. Whether we get this done is not dependent on European Union bureaucrats, not on United Nations diplomats and not on the Standing Committee of the National People's Congress of China. We can just go ahead and do it.

We already *have* a planet of plenty. The idea then that ten, fifty, even a hundred billion people can live on earth with *more than enough* is far from far-fetched, it is eminently attainable. And no, the fact alone that something is attainable does not make it necessarily desirable, let alone necessary. That is yet another question we'll want to ask ourselves: what do we want to attain? And so what we're saying with this book; our statement, if you like, is not: 'we can, therefore we have to,' our statement is: 'we can, therefore we can relax.'

I   THE ALTERED STATE

So what has changed, why is this possible now, when clearly it wasn't before?

We are beginning to gain an understanding of some seminally significant developments – shifts, you could say – that are taking place in the energy sector, as elsewhere.

The most important of these is that information technology is taking a primary role now in every aspect of our lives, including energy. In fact, energy technology and information technology are fusing together, and this has extremely far-reaching consequences for everything we do with energy: how we 'produce' or garner it, how we distribute, channel and trade in it. Crucially, it means that some of the principles that apply to information are beginning to apply to energy. 'Kilobytes as kilowatts' might be the catchphrase. And the most startling, also most potent, parallel between kilobytes and kilowatts, and one which is entirely new, is this: the more we use them, the cheaper they become. With kilobytes we recognise this to be the case immediately: Moore's Law,[IX] which we will explore in more detail a little later, postulated exponential growth for data processing as early as 1965, and it has been proved essentially correct ever since. So if, for the first time ever, the same is true now also for *energy,* then that means we are dealing with something we don't need to hesitate calling a genuine game changer.

The more sceptically minded among you may well think 'we've heard this kind of thing before' and detect a whiff of 'too good to be true' about it all. And while we don't think that it is too good to be true, we also don't think that anything quite like this has happened before. Nor do we think that it is simple or straightforward. If it were, we would hardly feel the urge to write a whole book about it, we could sum it up in two or three easy take-home statements and be done with it. The reason we feel compelled to write this book and go to some considerable lengths to explain the ramifications, the technological workings and indeed the wider philosophical context of this development, is that to our minds they merit proper examination and actual understanding. And this, in our own experience, just doesn't happen over a soundbite or two. But it does happen through some in-depth engagement with the changes that are now underway.

And so from this perhaps stems the 'big ask' of this book. We are not going to end on a call to action or a plea for you to sign up to some campaign, nor are we going to encourage you to buy any product, write to your political representative or take to the streets. We are not even going to ask you to change your behaviour or do something positive, whatever it may be. We simply ask that you play the thoughts through in your mind and stay with us for a while. It's going to take a bit of your time, and we will do our best not waste it. But our main ambition is to instigate and help usher in a different way of thinking, and we know, again from our own experience, that this takes a bit of effort. Letting go of our familiar patterns and saying goodbye to truths we have got used to like old friends is a wrench. But we think it's worth it, and we hope that by the last pages of this book you will agree that it is too. You may not agree with much else, and that is also just as it should be: we are not looking for a consensus, we are looking for a stimulating debate, because that, a really good debate and an open mind, we think, stands a decent chance of getting us somewhere worthwhile.

II   SEIZING THE MOMENT

For us, the energy question as it stands today presents not a problem or a challenge so much as an opportunity. An enabling moment in our own history. What we are witnessing right now can be blown up to epoch-defining proportions, or played down to an inconsequential blip, it mostly just depends on your own take on the matter. And to us that in itself isn't even of such great interest, because time will surely tell. We are not setting out, here, to predict how things will pan out, nor is our main concern to impress on everybody what an immense milestone we're at. What we have done with this book is actually simpler and more straightforward than that. We have recognised – as have many others – that as a species we are facing a pressing set of circumstances, and we have formulated our thoughts on what this means for us in both practical and theoretical terms.

     Within this, the biggest and most important point we want to make and examine is that as of about now, energy – how much we have, how much we use, how we use it and what for – has nothing to do with resources. The energy question on planet earth is shifting from one of how many resources we have to one of how we, as human beings, use technology to handle energy in principle. It is a question of logistics: how we get the energy from where it is to where it is needed, with minimum hazard. And that means it's a question of how we use our brains. Because logistics is organisation and that's something we can do in our heads, and if we can do it in our heads, we can implement it in practice. Energy, to-day, really, has to do entirely with our own ingenuity and how we apply this ingenuity to our home planet, earth. And *that* is why we call our book *A Genius Planet*.

THE TASK IN HAND
II

There are no serious predictions that the world population is going to go on growing indefinitely. Most forecasts – by the United Nations,[X] for example – have us stop just short of 10 billion by 2050 and then reach 11.2 billion by 2100. Some predictions – often labelled 'pessimistic' – draw a spectre of maybe 15 or 20 billion, maximum.

But let's toy with the idea anyway, out of curiosity. Say the earth's human population were to reach ten billion by the 2050s and then double by the end of the century to about 20 billion. And then again, over the next fifty years or so to 40 billion. And then again.

At what point would we 'tip the balance' and perish? What if we were to expand further, to two hundred billion, five hundred billion, even. What about 1,000 billion: a trillion human beings, on earth? Nothing about the way we currently experience and understand economics, ecology, agriculture and energy management teaches us that this would be even vaguely feasible. Resources would have been depleted, our home planet would have become a desert wasteland long before we would ever reach any such number, and we would simply have multiplied ourselves to extinction. It would not just be an unmitigated disaster, it would simply be impossible.

Or would it? Of course, if we assume that we'll just continue doing everything in exactly the same way as we have been over the last few hundred, maybe few thousand years, then the thought of even relatively minor population growth on the one hand and of giving everybody in the world comparable living standards on the other is nothing short of frightening.

But we are *human beings!* We are not going to continue doing things in exactly the same way as we have done over the last few hundred, or few thousand years. In fact, there is no single way in which we have been doing things over any length of time, really. True, sometimes it feels like the changes we're making are slow and cumbersome, and in many respects it takes us a long long time to adjust, but the reality of the matter is that we've been developing, inventing and evolving incessantly, since the word go. Our whole existence today is characterised by marvels we ourselves have brought about, each one amazing in its own right. And we use the word 'marvel' here deliberately: we habitually do things every minute of every day that to people only a couple of generations ago would have seemed imponderable. And they in turn did things their own forebears could not have dreamt of.

Even the notion of there being ten billion people living on planet earth at the same time would have been, to those who were living here two and a half thousand years ago, unimaginable. Around the time of Socrates, there were living on earth some 100 million people.[XI] Had somebody suggested to him that in 2500 years our species would multiply a hundred fold and still thrive, he would probably have thought them mentally untrustworthy, though he may have been pleased to hear that we still revere him as one of the fathers of our civilisation. Is it surprising that to us another 100-fold increase seems wholly inconceivable? Of course not. Is it *actually* inconceivable? Probably not.

We are obsessed with numbers, and there are good reasons for this too. As we shall see later in this book, we've invented the numbers that quantify our world, that therefore underpin our understanding of it. So it need not surprise us if we treat them as markers of reality. But is it possible to imagine a world in which the numbers – the symbols of quantitative dimensions – of one era simply fade into insignificance as they are replaced by another? Yes it is. We've done this many a time, in all manner of contexts. And as we develop new sets of numbers with which to describe our world, we shape it with ever increasing complexity and sophistication, and at new orders of magnitude. Today, we routinely build towers that accommodate whole towns, that would, two hundred years ago, have *had to* fall down. We have in the air, at any given time, the population of a small city – between about 400,000 and half a million people [XII] and to us this doesn't just sound perfectly reasonable, we shrug it off as un-noteworthy. It wouldn't have been un-noteworthy to Charles Lindbergh,[XIII] when he made the first transatlantic solo flight from New York to Paris in May 1927, less than a hundred years ago.

Once a shift from one technological plateau onto another has taken place, the *quantities* are no longer what interest us. What interests us is the *qualities* that they entail. We are not surprised to find that we can nip across the Atlantic by plane, nor are we impressed by it. That's just something we do. What interests us is how comfortable our seats are, what the food tastes like, how good is the selection of drinks on offer and what's available on the inflight entertainment system. And we certainly don't want to pay over the odds for it all...

When it comes to energy, the reason we are hung up on population numbers and the resources these numbers consume is because we are used to thinking in terms of resources and 'consuming' them in the first place. No sooner do we realise that that's no longer necessary than we also realise that these numbers become – quite spectacularly – irrelevant, and with this book we will be able to show how and why this is so.

I   **THE RED HERRING: SUSTAINABILITY**

Much of the energy debate since the 1970s has been focusing on exactly the wrong thing. Most people, quite understandably, concentrate on how we can make our energy use sustainable, by which they mean how can we use and manage our existing resources in such a way that they don't run out. It's not a silly endeavour. As long as you think of energy as being tied up with resources, and resources either getting scarce or damaging the environment when being extracted or used, it makes perfect sense. Sustainability is a serious, worthy cause, as long as you think that you have an energy problem.

But we do not have an energy problem. We don't have a resources problem either. There are plenty of resources. We don't have a land problem and we don't have a water problem. Our planet is literally awash with water, we use a fraction of the land that's available and we are bathed in energy.

Our problem isn't even one of the imagination. We are clearly able to imagine and re-imagine things, even things which are, at the time they're being imagined, so far fetched as to be, supposedly, 'unimaginable'. No, our problem appears to be largely one of perception. We have allowed ourselves to be panicked into a state of thinking that we have a crisis. And so we act as people mostly do when they are panicked: we make rash but timid decisions and limit our horizon to what we see as the oncoming headlights of a juggernaut. But we're actually already well underway to finding extremely effective, constructive and infinitely replicable ways of utilising the immense surplus of energy, space and resources we have, of managing it and making it accessible to everyone.

II   A CATEGORICAL LEAP

That's our challenge, our task: to manage the abundance of energy we have and make it accessible to everyone. And we can do this. We can do it, led not by ideology or doctrine but by an imaginative pragmatism that allows for a genuine competition among technologies and utilises a multitude of technology applications across the globe. And it really won't require that much from us.

It would be easy and satisfyingly dramatic to paint a picture now of a world on the brink of catastrophe, whose only salvation lies in a complete overhaul of everything we've ever known. But that isn't necessary. What will see us into the next chapter of our energy story, and what is already underway, is a gradual, incremental series of adjustments. What has to change *radically,* and soon, is our thinking, our approach. Because without that, these gradual, incremental adjustments, which are in themselves quite easy and very manageable, won't happen, and if they don't happen then we get stuck with our resources and sustainability issues, and then we really do have a problem.

So what you will get with this book is a new take on how we can understand energy and how, therefore, we can deal with it. What you will not get – and we know that you may find this frustrating – is many a telling statistic or table of figures. The reason for this is that we don't, with this book, want to concentrate on, and obsess about, numbers. Does this mean we are going to just ignore the facts and make it all up as we go along? Absolutely not. But not only do we think that the figures have been covered and can be looked up, discussed and verified or disputed elsewhere, we also think that working out projections on the basis of what we've been doing so far, what we already know and what we're therefore used to, and then making policy within those parameters is simply not going to get us anywhere near far enough. Rather, it's keeping us stuck in the precise same mindset that has got us into our situation in the first place. It was the great Albert Einstein [XIV] who said, "we can't solve problems by using the same kind of thinking we used when we created them." [XV] And we agree. Which is also why we think that nobody has really quite put their finger on a way forward yet.

In the process of researching and writing this book, we've done a lot of travelling around, holding lectures and seminars, attending conferences, all over the world. And time and again, when we are talking to people – often very well informed people – about this subject, they almost immediately default to one of two or three stock responses: 'it's not going to work, renewables only make up a tiny proportion of our energy provision'; 'it's not going to work, the established interests in the energy industry will never sit back and let this happen'; 'it's not going to work, it costs far too much.' All of which may be either true or at the very least debatable in the traditional framework that has been applicable for the last few hundred years, more or less since the Industrial Revolution. None of which though will be applicable in the emerging framework, as we shall be able to explain.

So we will reiterate: you have to read this book with an open mind. If you come to it with your mind made up as to what's 'possible' and what isn't, what 'will happen' and what won't, what 'the system can take' and what it can't, then not only will this book not inspire you, it may positively annoy you, because it won't deliver what you are looking for: a rephrasing of an outdated paradigm. (We've been told to avoid the term 'paradigm', because its currency has been somewhat devalued through overuse, but in this instance it's merited, because it is accurate.)

The very reason we have written this book is that we find what is still missing from the debate, and what nobody has really proposed in earnest yet, is a *categorical leap*. A genuinely, *fundamentally*, new approach to energy that, through being implemented, lifts us onto a different plateau altogether. And yes, it has to be implementable, it has to be founded in practically applicable technology and supported by a sound theoretical framework. We think we have got that and are able now to present it with this book.

III   THE CURRENT CONTEXT

Clearly, we are not the first people, nor are we the only ones, who work towards this particular goal. Since we started working on this book at the beginning of 2009, a lot has happened in the world of energy, and there has been a noticeable shift from a fixation on the 'crisis' to an outlook towards 'solutions'. Robert Galvin [XVI] with The Galvin Electricity Initiative and his book *Perfect Power,* [XVII] talks about very similar things as we do, in a very comparable vein. (Our book was even going to be called *The [Power] Book* for a while.) Galvin, though, takes an exclusively US-focused approach, and he stops short of addressing the big, overriding picture on a global, let alone conceptual level. What Bob Galvin has in common with some of the other recent contributors to the debate is that he talks about a power 'revolution', which will wrench us off fossil fuels and bring about a safe and sustainable energy future. Carl-A. Fechner, the German documentary film maker who directed and produced *The 4th Revolution* [XVIII] does pretty much the same thing. But while he takes a global perspective, he does not acknowledge the extraordinary potential that lies in networked, digital technology and talks much more about localised solutions. Jeremy Rifkin, with *The Third Industrial Revolution,* [XIX] goes further and recognises the power of network technology. But he treats it, much as the title of his book suggests, in the context of existing power structures and thought patterns and consequently concentrates on a master plan approach to the implementation of a grand scheme.

Although we think that neither of these contributions go far enough, nor really address the conceptual basis for what is now happening and what is possible in the world of energy, we do think that they are all welcome in so far as they steer the discourse about energy in a different direction. And the general thrust of that direction is by now fairly clear: away from fossils. Notwithstanding any advances that may or may not be made in carbon capture or shale gas extraction technology, there really are hardly any solid arguments for thinking in terms of coal, oil and natural gas any more in the long term.

There is, however, a big and important differentiation to be made between what we see as the long term development arc and the short to medium term stages on the way there. Sometimes these are viewed as if they were two opposite ends of a spectrum and discussed as either/or choices: either you have fuel powered cars and everything keeps running, or you have electric cars and you need a complete immediate overhaul of the entire road transport infrastructure. But in actual fact, they are part of the same story, and what matters is getting the right kind of an and/and solution, that may, or may not, be a transitional phase, for example using excess wind power or solar energy to make clean synthetic fuels, which can be used for as long as the existing infrastructure is in place while being adapted, no matter how long that process then lasts.

Towards the end of the first decade of the twenty-first century it looked as if something of a consensus might emerge on climate change and carbon emissions. A move away from fossil fuels, more and more people agreed, would be a good idea. Yet as we write this, in 2012 [and as we revise it and prepare it for publication in 2016], there is still as much disagreement on virtually every aspect of the energy and climate debate as there has ever been. For every person who talks to us about melting ice caps, we meet someone who tells us that oil will be plentiful in supply and cheap to burn for decades to come. And even if we accept that right at the moment a really very significant proportion of the scientific community interprets climate change and global warming as man-made, from there on in everything is pretty much up for grabs.

That was the case before 11th March 2011, and since then the questions have only been magnified. With the most serious nuclear incident in a quarter of a century unfolding at the Fukushima [XX] plant in Japan, the risks and environmental costs of nuclear power were brought right to the fore of people's and government's minds, to the extent where Germany, Europe's largest economy, decided do terminate its atomic energy programme within a decade. (A policy that has since also been adopted by Germany's much smaller neighbour to the south, Switzerland.) Meanwhile, large-scale coal fired power stations continue to be built on a weekly basis, and the discussion about carbon capture, carbon emissions and indeed the level to which our carbon emissions either contribute to or are responsible for climate change rumbles on.

So, on a global scale and over the long term, what are we going to do? Are we going to go nuclear after all? Are we going to go decentralised local? Are we going to go super-sized solar? Are we going to have a cocktail of energy sources? Are we tending towards micro-generation? Will that suffice? What will it look like? Are we going to cover the countryside with wind farms? With solar panels? With more and more pylons everywhere? What role, in all this, are smart grids going to play? And what, pray, is all of this going to cost? Can we afford to do all of this? Can we afford not to?...

IV   WHO WE ARE

We have something to say about this. And by asserting that we have something to say about it, we are, of course, planting ourselves right in the middle of one of the most important, and therefore also most controversial, discussions that are currently taking place. And that is just as well, because while we are not on the surface the obvious narrow experts on the matter, we are knowledgable about the technologies and the theoretical framework that now come into play, and so we can look at this issue from a different angle. And that, we believe, is exactly what's needed: a fresh perspective.

So who are we?

Ludger Hovestadt is professor of Computer Aided Architectural Design (CAAD) at ETH Zürich.[XXI] ETH *(Eidgenössische Technische Hochschule)* Zürich, the Swiss Federal Institute of Technology, is one of the most highly regarded technical universities in the world. As an architect, Ludger is particularly interested in spaces, and how people interact with them. If this sounds slightly obvious, then that's because it is. What isn't entirely obvious, though, is how you get from being an architect to talking about energy. In Ludger's case, this has to do with two pertinent facts: firstly, Ludger is also a computer scientist and, bringing the two disciplines together, he has a significant body of experience in developing, and heading up research into, 'intelligent' buildings. Beyond the design of buildings, one of Ludger's main areas of interest is how to make the complex infrastructures that permeate our daily existence not just usable but user-friendly. Crucially, out of this fusion of interests has come an invention, the digitalSTROM chip. We will be explaining this a bit further on, but, in a nutshell, it offers a turnkey element to making smart grids, micro-generation and indeed in the longer term an 'internet of energy' a possible reality.

Vera Bühlmann is a co-founder and was until 2016 Head of the Laboratory for Applied Virtuality at Ludger's Chair for CAAD. She holds a PhD in philosophy from Basel University and her background lies in cultural theory. What interests her is artificiality in general, and within this, the role that technology plays as 'the explication of metaphysical ideas' more specifically. One of her greatest concerns is the increasing level of inarticulacy between disciplines, particularly at a time in history when six out of seven billion people enjoy basic education at least to the level of literacy. To her, "some of the most basic notions constitutive for cultural forms of domestication, such as those of information, of unit and value, of formula, code and formalisation, today lack a reflection – and hence a consideration and awareness – of the acts of abstraction each of those notions have come to integrate and incorporate symbolically, throughout history, that is complex enough, and therefore adequate, to meet the unprecedented performance of contemporary technology at eye level." With energy being so much at the core of everything that exists, she sees as maybe the most profound challenge today a rethink of what we consider real, possible, feasible, or reasonable. And this not just for the sake of it, but because our perception of what we can do, indeed the very directions into which we are able to look, are conditioned by our intellectual abilities, once a basic level of literacy has been mastered. We ought to, she feels, address these issues anew, not from the point of view of what we may consider possible in an absolute sense, but in view of the wealth or enrichment they brought us throughout the history of intellectuality. In order to reappraise the notions of our world's reality, we can invoke an ancient term that has recently earned itself a great, if somewhat dubious, topicality: the 'virtual'. Starting September 2016, Vera is Professor for Architecture Theory and Philosophy of Technics at the Institute for Architectural Sciences, Technical University (TU) Vienna.[XXII]

Both Ludger and Vera are therefore experts on how digital technology transforms the way we live, and in fact, during the last phase of writing this book – for most of the year 2012 – they were based in Singapore where they were conducting research as part of the ETH Future Cities Laboratory.[XXIII]

The confluence of these backgrounds – computer science combined with architecture on the one hand and philosophy in the context of the virtual on the other – makes our approach unusual, so as not to say unique (although it is probably both). Because while this book will talk in some detail about the technological aspects of our proposed way forward, it is rooted in a comprehensive theoretical foundation. It's a foundation that is in places quite complex, and so in order to keep this book accessible, and to let it flow without too many detours into abstraction, we in places rather skim the theory. But we will, of course, be giving references to further reading in the Appendix at the back.

Sebastian Michael, meanwhile, is a writer of several stage plays, a novel, two musicals, an ongoing conceptual narrative; as well as two shorts and a feature film, which he also directed. "I think, write and create across disciplines in theatre, film, video, print and online with a deepening interest in humans, the multiverse and quantum philosophy," is how he describes his work, and he is, as Peter Brook once put it, "interested in the universe and everything that surrounds it," which may well be a superverse of all possible multiverses... Sebastian views our human existence – and our own awareness of it – on this particular speck of cosmic dust as the most wondrous, intriguing, fascinating and engrossing phenomenon at every level imaginable, be it bio-physical, emotional, inter-relational, theoretical or even conceptually spiritual, and so for him, finding a language to talk about it is the great personal challenge. He has no scientific background whatsoever and that is precisely the reason he has been called upon to team up with Ludger and Vera to put our thoughts into words that a lay person like himself can understand.[XXIV]

V   WHERE WE'RE COMING FROM...

Ludger and Vera decided to work together and get Sebastian to write a book about energy because they felt they recognised a development in their world that fascinated them and that they each understood from their own field of expertise, but that did not seem to have been brought into relation with energy anywhere near enough in the broader discussion, even though it is directly and extremely relevant: the move towards ubiquitous computing.[XXV]

How does this manifest itself? Since about the 1950s, when large companies and government departments started spending heavily on information technology, it had become clear to anyone working in the IT field that without it soon nothing would go. But hardly anybody else thought so at first. For the majority of consumers, computing was not only not an issue in their day to day lives, it just didn't feature at all. For a while. But in the second half of the 20th century, we saw big main frame computers being succeeded very quickly by personal computers, and no sooner had we got used to the idea of everybody being able to own their own PC, than these machines became portable and we all started carrying around laptops. By the 1990s, these many individual computers started talking to each other over the internet, and as a natural consequence of things being shared and done collaboratively, we began to rely less on applications that were resident on our machines, and instead reached for 'the cloud' – servers around the globe connected to us and each other over the net – to do our computing.

While all this has been happening, more and more of our electrical devices have become equipped with some sort of processing power. They no longer just have an on/off switch and some analogue regulation of their performance, they do things in a programmed way. Often the way they do this is still very simple, but it is nevertheless programmed. Computing, once confined to large rooms in massive buildings for the purpose of some highly specialised tasks, is now everywhere, and we use it for everything without giving it a second thought. The average smartphone in our pocket holds considerably more computing power than the control centres that guided manned missions into space back in the 1960s.[XXVI]

The fact that we have computing everywhere and in everything means that we now have an abundance or saturation of computing. As a result, we have ceased thinking about computing itself, and instead think very much in terms of how we orchestrate different instances of computing: how my laptop works with my smartphone and integrates with the network at the work place, for example. We no longer think of tasks and invent machines that can do these tasks, we have advanced to a stage where we have devices and let ourselves be inspired into inventing uses for these devices, simply because we can.

This process reaches much further than what we associate with 'IT', because it has long since taken in not just devices that we have for the purpose of handling information, but devices that handle information for the purpose of doing something completely different, such as washing machines, or air conditioning units. And the same thing also applies to energy generating devices, such as solar foils, wind turbines and geothermal installations. All these machines, installations and devices are now beginning to be *orchestrated:* made to work together, integrated and connected.

Because computing has become ubiquitous, we have seen what we might call a 'secularisation' of computing: everybody uses computers, not just the initiated. Computing has long ceased to be the rarefied domain of experts. We all can – you might say we all have to, to some extent – engage with information technology.

The effect that this has on the one hand is a baseline democratisation: computing belongs to everybody and everybody can do with it what they want. But on the other hand it also has the effect that we now tend to use an extremely potent and sophisticated technology in really quite a rudimentary way. This is by no means unusual: anything that is thus 'secularised' – taken out of the hands of the expert few and rendered into the hands of the inexpert many – will, in a sense, flatten, as users themselves make basic use of what they have at their disposal, while at the same time programmers and designers latch on to the need for creating applications and devices that can be learnt from scratch in an instant. User-friendliness and intuitive interfaces are among the chief characteristics of ubiquitous computing.

This 'democratisation' is not a bad thing, but by using computing in a rudimentary way, we are missing out on the great, complex potentiality it offers, and in the context of energy, this failing makes itself particularly felt, because energy is so important to us. That's why Ludger decided to take a step back from the practical application of information technology and join forces with Vera who could offer a theoretical perspective: they both felt that what we really could do with was a new take on this whole issue, and a new sophistication. They realised that in their constellation they had a great opportunity for communicating not just a practical set of information – a 'how to' of solving the 'energy crisis' – but also the theoretical framework – the 'why' – for doing so.

Put in the mix Sebastian's background as a writer, storyteller and communicator, and we are able, we think possibly for the first time, to fuse a technical perspective with a philosophical one and put it in words that are not necessarily simple, but that we hope can be followed and understood. This matters a great deal to us, because we get a strong sense that some of the themes we are talking about in a theoretical or philosophical context have started to feel so lofty over the last few decades, and at the same time the vocabulary for them has become so inadequate, that many people have effectively switched off and now regard them as tangential or irrelevant, when in fact they are central to our understanding of who we are and where we're at when it comes to energy.

You will hear us use the phrase 'our take on energy', because this – the way we look at energy, how we deal with it, handle it, trade it, how we think about it, how it integrates with our lives: our whole understanding of what energy is and what it means to us – lies at the heart of the energy 'issue', and our take on energy is, at its core, centuries old. In fact it dates back several millennia. This is not something we have to admonish ourselves for, it's just the way it is, it's how we, and our concept of energy, have evolved. There is nothing wrong with it. Or rather: there *was* nothing wrong with it, until recently.

Up until very recently, our ancient, habitual way of doing things was perfectly adequate, and it entailed basically burning fuels that we find in, or grow on, the ground to generate heat. By and by, from heat we learnt to produce motion, but the source of our energy was still the same: fuels we find in, or grow on, the ground. In the process, as everybody by now is acutely aware, we release a plethora of particles and gases, and so the more we burn, the more we put a strain not only on our resources, which as a result start to dwindle, but also on the rest of the ecosystem, because we're overburdening it with substances that aren't normally there in the form we're putting them there. And this is what we've been doing for ten thousand years. It is fair to say that the principle at work in a big coal-fired power station that produces maybe one or two gigawatts of electricity and supplies a few hundred thousand people with the energy they need to get by is not in essence any different from the principle that was at work when caveman gathered up some logs and built a fire to keep his family warm and cook dinner.

It's a principle that has worked fine and got us far, but it's not going to get us much further. Why? Because there is only so much 'stuff' in the ground. It doesn't matter what the statistics and the forecasts say, whether they predict that coal will last another hundred years or two hundred, that oil will last another fifty years or seventy-five... – the figures will come from somewhere and they will make some sense. Someone will be able to prove that it is so, and someone else will be able to use a similar calculation to prove that it isn't.[XXVII] What doesn't change is the principal fact that this 'stuff' in the ground is limited, finite. One day it will have gone and that will be it. The amount of time it will take for it to replenish features on a scale so far off our own frame of reference in the context of our existence on this planet that we may discount it as irrelevant, which means that as far as we are concerned, finite resources are finite. And there are only so many trees and crops we can plant for fuel, not least because we'll also still want to grow food to eat and have room to live. Yes, there are renewables, such as biofuels, but no, they are not the answer. The reason they are not the answer is that they are incredibly inefficient at what they do, which is converting one type of energy (solar) into another (combustible mass), before we come along and chop them down and burn them, thus converting the energy once again, this time from combustible mass into heat and maybe motion and maybe electricity. There are simpler, cheaper, more effective ways to turn solar power into heat, motion and electricity, and we'll be talking about them in some detail.

We have available to us today technologies that allow us to wean ourselves off fossil fuels entirely. And that means we don't have to rely on this millennia-old principle any longer. We can cover our energy needs from an endless, inexhaustible source of energy: the sun. The sun is the source of almost all our power on this planet anyway, but while up until now it has been necessary for us to detour via natural resources, we are now in a position to access a lot of this energy directly. So much of it, that we do not have to worry much about the rest. You may have heard this before: the sun expends on planet earth around 10,000 times as much energy as we are currently using. So even if our demand grows, substantially, over the next few hundred years, we will still have plenty. What we need to do is find a way of getting at this energy, and making it useful to us. And there is: we are talking mostly about photovoltaic solar power, combined with some other forms of, not 'renewable', but perennially available sources of energy, namely geothermal, thermo-solar, hydropower and wind. Add to this solar fuels, and even the 5 %-7 % of our energy consumption that until very recently looked like the hardest nut to crack – fuels for aviation and shipping – move within reach of a useful, solar-powered alternative.

Understand us correctly: we, as the authors, are not fussed about solar, or any other type of technology, in particular. We don't mean to say: 'here is one solution and one solution only'; what we are saying is, 'here is a way of creating a palette of solutions; they are entirely viable and absolutely worth considering and pursuing because they are what's going to get us not just a little further down the alley, but onto a whole different kind of avenue altogether'.

And let us assure you: we are not making this up. Nor is it wishful thinking that bears no relation to reality. Yes, there is a concept behind it, there is a theoretical background to it, but it doesn't stand alone, and it isn't what drives the thinking. What drives the thinking is a simple imperative: we need solutions – genuine, long-term – to our predicament. And we have at our disposal technology that can offer up such solutions. This technology sits right within the development framework that we see in action to astonishing effect in other areas, namely mobile telephony and information technology: networks becoming the standard form by which we organise ourselves and what we do. We will be talking a fair bit about this, and therefore we won't go into any detail just yet, but the significance this has, for us as a species, is phenomenal. Because not only are there new types of behaviour emerging, and new power constellations as well as new economic platforms, every day, even as we type this, but we are only just at the very beginning of realising the potential this entails. So if, throughout the book, you get a sense that we are perhaps labouring this point a bit, the reason we're doing so is because it merits a lot of our attention. And it merits our attention because it is network technology that emancipates us from our old way of resorting to earth and its substances for our energy needs. It is *network technology* that enables this shift from a resources-based energy approach to one that is logistics-based.

None of this is completely new. We've already started the process; in fact, as we'll be able to explain, we're a good four or five generations into it. Because 'network technology' is nothing more and nothing less than an extension of information technology, which in turn is an extension of electricity. So while this may sound like a contradiction, it really isn't: we are quite fundamentally changing our relationship with energy, and we already know how to do it. What we haven't really done yet is adjust our thinking, our mindset, if you like. And it would be better if we did adjust our thinking, because we're in danger right at the moment of taking these new technologies and instead of using them to their enormous potential, continue to use them the way we've always used energy. But that wouldn't make any sense, because, as we've realised, the way we've always used energy isn't going to get us much further now. It's not going to cater for the extra billions of people we're expecting on the planet, and it's not going to solve our resources related and environmental problems.

This is why we will also be talking a fair bit about making use of our imagination, about encouraging invention and not setting our sights too low. Because what will get us out of our predicament is visionary thinking and invention. It's setting our sights far, high and wide, beyond our current horizon; it's turning our focus away from the abyss and the impending calamity of stepping into it, and instead looking up, ahead to where we may get to, if, rather than just keep on running, we take off and fly...

So the place we are coming from with this book is one of optimism. It's a word that hadn't been heard in the context of energy for a fair old while, and this in parts is exactly what had started to grate on us. Now, it's slowly coming back: the voices of doom that had been dominating the literal and metaphorical airwaves for about a decade or so have begun to give way again to the spirit of looking up, looking forward. We like that. And this is not even so much about 'progress' – in itself a contentious term – or about making everything 'good', nor let alone about solving all the world's problems in one fell swoop. What this is about is taking a situation and dealing with it in a constructive manner so that things can move forward and in the process get better. And things can and do get better through our input. If you happen to think that the opposite is the case, that we human beings only make things worse, then stay with us. We will be able to illustrate to you that you have reason to have faith in your own species, even if some of what it does does at times look wantonly destructive.

But just before we move on to explain the constructive manner in which we envisage us all together coming out of our 'energy crisis', we are going to briefly, and without dwelling on it, dip into the negative side, the things that annoy us, just to complete the picture and so you know where we're coming from, from all angles.

One of the most memorable sequences in the 2006 Academy Award winning documentary *An Inconvenient Truth* [XXVIII] has Al Gore step on a fork lift crane to reach up to the predicted level of CO_2 concentration in the atmosphere by 2050, as the corresponding graph, which has been shown to go up and down in a regular pattern over 600,000 years, fluctuating between about 30 and just under 300 parts per million, shoots right off the scale to well above 600 parts per million. It's a startling image to see the man who 'used to be the next President of the United States' whirr up into the air to predict the dire consequences of what would happen if this scenario should be allowed to come about.

The book that accompanies the film makes the same point in almost as compelling a fashion, by means of a flap on the page which folds out to reveal, once opened, the near vertical incline this curve takes, as it leaps from today's already worrying 370 ppm to the cataclysmic 600+ ppm anticipated for the middle of this century. The words Al Gore uses in his book to describe the attendant anxiety are:

"There is not a single part of this graph – no fact, date, or number – that is controversial in any way or in dispute by anybody. To the extent that there is a controversy at all, it is that a few people in some of the less responsible coal, oil and utility companies say, 'So what? That's not going to cause any problem.' But if we allow this to happen, it would be deeply and unforgivably immoral. It would condemn coming generations to a catastrophically diminished future." XXIX

The message is clear, stark and twofold:

1)   There is a truth – it is undisputed and indisputable, and anyone who says otherwise is irresponsible and only does so because they have vested interests
2)   The truth is that unless we act swiftly to stop a particular development or phenomenon from occurring, we will condemn future generations to catastrophe

'Nothing', indeed, 'is scarier than The Truth', just as the tagline of the film told us. What was really disturbing though, and what in all seriousness could be doing untold damage to our world right now, because it obfuscates science, distracts us from clear-headed thinking and depresses us into inaction at a time when we need inspired invention, enlightened polity and decisive but diligent action, is the apocalyptic tone that had taken hold.

Small wonder then that we started to feel, and behave, like lemmings. Instead of taking charge of the situation and dreaming on a visionary scale, we delegated our decision making to politicians and scientists: authority that we can't entirely have confidence in, but that we can at least defer to and say, 'It's out of our hands.' But politicians are as scared of us as we are of the imminent disaster. They have lobbyists breathing down their necks, industries pressurising them, and votes to lose. Meanwhile, scientists started sacrificing objectivity for security. So rather than say that they disagree with 'widely held views' or contradict that hitherto most elusive, because least probable, of outcomes, the 'scientific consensus', they exercised self-censorship, because if they didn't, they risked losing their funding, and if they lost their funding, they were no longer able to carry out their important research.

Scientists who censor themselves for fear of being seen as heretics? Truths that are not just inconvenient but also incontrovertible? Politicians under pressure to come up with simple solutions to complex issues, just to appease the masses? There's a distinct whiff of the dark ages about this. The zeal and conviction with which we are being berated to curb our carbon emissions, and the categorical *verbot* of any dissent, whether it's scientific or not, are more characteristic of a totalitarian ideological dictatorship or a religious autocracy than of a global community of 21st century pluralist democracies.

And so we find ourselves paralysed, incapable of moving forward because in front of us we see an unfathomable chasm, and behind us we're being chased by escalating statistics of dizzying heights. The most we can do, it appears, although we're far from convinced that it really makes any difference, is take the stairs instead of the lift and turn the telly off standby before going to bed in order to reduce our carbon footprint and, more importantly, to soothe our conscience. Don't get us wrong: taking the stairs instead of the lift is good: it's healthy exercise. But it's not what's going to get us onto the next floor as a species. Whereas looking past the cloud in the sky, beyond the silver lining, adjusting our horizon and coming up with new ideas, is.

And that very nearly, and neatly, brings us to how we want to approach things. But there's another point we need to get straight, and straight away. And that is this: we are not 'green'. We are not pro-this or anti-that. We are not environmentalists and we are not 'in cahoots' with anybody. Ludger, as we've already pointed out, is a professor with a permanent chair at ETH Zürich, and he's an inventor. He's also an entrepreneur, and this goes hand in hand with his academic and development work, from which several spin-off companies have resulted. One of the companies he helped set up was Aizo, with which he developed the digitalSTROM chip, and he is a co-founder of the digitalSTROM Alliance, originally set up at ETH Zürich to advance the technology to a global standard. Since then, Aizo has been sold and Ludger is no longer involved with digitalSTROM as a brand, company or product.[XXX] We will nevertheless be talking a fair bit about digitalSTROM, because as its co-inventor Ludger knows the technology inside out, and it serves to illustrate some of the things that are possible, but we want to make it clear what we are about and what we are not about. We are not about the promotion of a particular technology, product or solution, we are not about a political agenda of any shade, and we are categorically not about any type of ideology. We are about a change in tone and perspective.

There are, on all sides of the spectrum, deeply held 'beliefs' as to what is the 'right' and what is the 'wrong' way of going about addressing the energy and CO_2 issues. Even just lumping them together in one sentence like this is viewed, again on all sides of the spectrum, as a declaration of some quasi faith: either you 'believe' in man-made global warming, or you don't. We don't like this way of thinking. It's not scientific, it's not rational and it's not ground-breaking. It's deeply old-fashioned, and it's misleading. In science, if there is such a thing as 'truth', it comes in the context of the laws and constructs that we have applied to it. It is neither absolute nor is it arbitrary. It is, at every level, open to scrutiny, query, doubt and refinement.

So we don't want to discuss things in a tone and on a level of received wisdoms and allegiances. If we are going to talk about nuclear power, as we will, we don't want to do this from a pro-nuclear or anti-nuclear stance. We are not interested so much in what is expected of us, as we are interested in a discourse during and from which new ideas and original approaches can emerge. What doesn't help any of us is if we all fall in with each other and start 'preaching' to our own 'converted'. Again, you notice the religious vocabulary, and the fact that we're putting it in adverted commas. Both for good reasons.

With this book, we seek an enlightened stance, and we invite you to join us in this endeavour. And that means by definition that it is not a stance so much as a motion. Because we cannot, on this subject or on any other, stand still. The developments that have taken place, the change in tone that we've already witnessed, the proliferation of published material that has occurred, just over the few years we've been working on this, have been nothing short of amazing. When we first started talking about this book, many of the ideas in it were so new that we barely came across them elsewhere. Now they have matured, they've gained currency, and they're being further developed. We're glad about that. It means we can bring them to a table that is already set, rather than having to assemble the table from scratch. Many of the ideas we talk about are also being talked about by other people, often simultaneously. Where we have consciously adopted something from somebody else, we will say so. If we don't credit anyone and you come across the same line of thinking elsewhere, it's not because we've nicked it from them or they from us: this is an area in which the ground is really shifting now. And that in itself is positive and most encouraging.

VI   ...AND WHERE WE WANT TO GO

Tom Lehrer,[XXXI] the American satirist, pianist, songwriter and mathematician, memorably observed: "It is a sobering thought that when Mozart was my age he had already been dead for two years." Wolfgang Amadeus Mozart[XXXII] died two months short of the age of 36. In Mozart's day, that was not at all unusual. Even as late as the early 20th century, the average life expectancy in Europe was just about that. The current world average is 71 years, and if you are born in Mozart's native Austria today, you can expect to live to over 80.[XXXIII] That's well over twice the number of years Amadeus had to write his 600-odd works.

Science, technology, and in no small measure energy have given us a world that Mozart might find exceptionally exciting, but simply wouldn't recognise. By the standards of our forefathers, we live in conditions closer to Paradise than most of them would have believed possible. Granted, not all of us are so fortunate. But more of us have better health, better education, more spare time, more disposable income; are forced to experience less physical pain; suffer lower risk of accidental or violent death, lower incidence of debilitating disease or loss of limb; enjoy more freedom of expression, more access to information and learning, more choice in pursuing our careers, interests and pastimes, fewer frustrated ambitions, less restriction by arbitrary authority and less limitation to our potential and self fulfilment than any generation before us.

As at 30th June 2015, Austria had a literacy level of 98.0 %. This put it on joint fourteenth rank with, among others, Serbia, Bosnia and Herzegovina, and Bermuda, and just marginally behind Switzerland, the United Kingdom and the United States, which, together with some two dozen other countries, perched at 99.0 %. In order to make the top 100, a country's literacy level, by 2014, had to be 95 %.[xxxiv]

In Europe around the 1770s, when Mozart was in his mid-teens, anything up to about half of the adult population were unable to read or write. The privilege of taking time off for yourself and simply do what you felt like for a couple of days at the end of every week and for a few weeks every year was reserved for the top percentile of the population. Going on holiday was, for the vast majority of people, simply not an option.

When Mozart travelled Europe in a horse-drawn carriage, with iron-clad wheels and bouncy suspension, being 'on the road' was dangerous, uncomfortable, slow and exhausting. If someone had told him, 'of course, you can board a train in Vienna in the morning and arrive in Paris in time for dinner; it's a good thirteen hour trip, but you'll have time to compose a concerto or two, take a nap, have lunch and see some of the scenery whizz by while you're at it', he would have thought you unsound of mind. 'Oh and from Paris to London it's only another couple of hours.' – 'There's a sea in between!' – 'It's all right you go straight under it through the tunnel, you won't even notice it's there...'

If you'd given Mozart a smartphone with his collected works on it, he would not have been able to believe his ears. What that thing does, to him would have suggested either magic or witchcraft. He would never have heard his music sound so clear, so pure, so perfectly produced in his entire short life. The idea that, while on a train from Vienna to London, he could not just write a piece of music, but orchestrate it, produce it, save it, mail it to his dad Leopold in Salzburg and, while still travelling, upload it to his website and share it with his fans all over the world, would have seemed so fantastical as to not be worth entertaining.

The cause of Mozart's death can not now be fully ascertained, but what is almost certain is that if he had been born two hundred years later, in 1956 instead of 1756, he would stand a very decent chance of taking his trip from Vienna to London today in 2016, at the age of 60, with many years of creativity and genius ahead of him yet. Public health care provision and contemporary medicine would very likely have cured him of the illness that afflicted and killed him, and there would be few obvious reasons why, having built on his reputation and begun to reap the financial rewards of his celebrity, he should not be doing really quite well. Have a look at Eric Clapton or Mick Jagger and you get an idea of the kind of lifestyle he might enjoy (except they are, as we write this, quite a bit older now, of course, than sixty...).

We have, no doubt, a lot to be worried about. But we have infinitely more to be proud of, to celebrate and to cherish. When your kids start school, the likelihood of them turning into healthy, fully rounded, well educated individuals who can exercise control over their lives and make practically all their important choices in accordance with their own priorities are fair to good. There are no guarantees, and the ecosystem on planet earth may yet collapse and prevent them from growing up at all, that's a possibility, it's true, but it's by no means likely. And we're doing so well as a species *because of* science and technology, not in spite of it. Technology does not by default restrict and restrain us: it can give us room to breathe, to move, to develop and to be. If you are in any doubt about this, ask yourself if you'd prefer your own little Amadeus to grow up now, or two hundred and fifty years ago, and you have your answer, right there...

All of which in a way illustrates where we want to get to: we want the extraordinary advances we've made in technology, the opportunities they bring us for genuinely improving our quality of life, but also the basis they create for us to do and be more as human beings, to become not just better off and have more convenient ways of getting about, but to be healthier, better educated, more capable of resolving our conflicts without resorting to violence, less in fear of harm and damage and more at liberty to pursue our interests and inclinations; we want these advances and advantages to flourish further and to take us onto the next plane, if you like, because that is now possible.

There is nothing mystical or 'New Age' about this. We may be inclined to call it the beginning of a new era, we're even prepared to give it a new name, but here, too, we want to caution against getting the wrong end of the stick. We don't see this as the Dawning of the Age of Aquarius. It is neither our ambition to bring about a happy epoch of bliss and harmony, nor do we think that there's anything otherworldly about where we want to go. In fact the opposite. We see this as another significant step in our development, brought about by all the previous ones, rooted in science and culture. We see neither necessity nor reason for any form of misology – the distrust of reason – and so this book certainly does not come from any spiritual perspective. We don't use words such as 'genius' or 'inspiration' in any sense other than relating directly to what we ourselves, through the use of our intelligence, our brain power, our experience and the intuition we gain from it – our ability to acquire, process and share understanding – are capable of.

VII   A PATHWAY OPEN TO EVERYONE

Everything has a cost to it. Financial, environmental, human. So the moment you come up against – or indeed with – the argument that we have to stop growth and expansion because of the environmental cost, ask the question, but at what human cost? If you hear it said that we can't afford development because of the financial cost, ask the same question: what about the human cost? Nothing is as lethal as poverty. And the only way out of poverty is development.

What we need is a path that is open to everyone, whether they're in Switzerland or in Swaziland, Iceland or India. And the kind of technology that we have in mind is indeed accessible, available and affordable to everyone, as our Case Studies a bit further down will show.

What we want to do is get to the next stage in our history, without leaving half the world's population behind, without stretching the planet beyond breaking point and without setting up now for generations to come problems that we aren't even capable of quantifying yet, let alone able to solve. No ideology is going to do that for us. What will determine the shape of our energy future more than anything else will be economics. By this we don't mean to say that pure economics is or indeed should be the only factor that counts. We don't hold with a world view that, as Oscar Wilde [xxxv] so poignantly put it, 'knows the price of everything and the value of nothing'. What we will have to do, as a society and as a community of global economies, is negotiate and define our values, while we are in our pursuit of energy, wealth and quality of life. And this means we will have to develop and evolve our ethics as much as we develop and evolve our technology.

The price we are willing to pay for risk-free, zero-emissions power may indeed in the short term be higher than what we are asked to pay for energy which carries with it a trail of associated environmental and economic costs while it is being generated and possibly long after it's been used. But this is not the way it's going to be forever. Before long, we will find that using our technological ingenuity to bypass natural resources and tapping directly into the energy that the sun and the earth provide will not be more expensive but in fact cheaper than using natural resources. There is a point in the very foreseeable future when photovoltaic solar, in particular, will be the cheapest way to garner electricity, and by some margin.

We are witnessing, already in progress, a move from a fuel-based energy model that is in essence industrial and hierarchically structured from the top down, to one that is digital and networked. And this is of great significance, because once we connect energy garnering and energy generating devices and installations, in some cases over fairly long distances, and make them part of an 'intelligent' network, it suddenly doesn't matter any more that the sun only shines in any given place for so long.

This has been one of two main stumbling blocks so far: the fact that solar power is often most abundant in places where it isn't strictly useful, or at times when it's least needed. So the first challenge is to get the energy from, say, Texas to New York or from the Gobi desert to Beijing. Or from your neighbour's house into yours, at no notice. The second challenge is to get our electricity grids to cope with a different way of feeding energy into it. Up until now, electricity grids have had to deal with about two dozen different types of power plant, and, depending on their size, maybe a few dozen to a few hundred energy-generating installations – for which read mostly power stations – per grid. And for the majority of these power stations it was comparatively predictable how much energy they were going to produce, and when. They could be *controlled.*

If we want to make use of our abundant sources of energy – the sun, the weather and the warmth in the earth – electricity grids will have to cope with *millions* of energy-generating devices and installations, from a few solar foils on someone's roof, to massive set-ups covering many square miles of desert, for example. This is inherently problematic, as the existing infrastructure is just not geared towards it. Power grids break down for relatively minor wobbles: they can just about handle fluctuations in energy consumption, what they can't handle very well yet is a volatile supply. So they either need updating or replacing. But building whole new infrastructures takes a long time and is prohibitively expensive. We can't expect people all over the world to readily buy into any solution that requires them to throw out their appliances and buy new ones, while rewiring their buildings, tearing up the roads to lay down new high voltage cables, and demanding of their governments and energy companies that they invest billions to accommodate a comprehensive new system, especially while the old one still seems to be working, even if only just. Nor can we expect developing countries to suddenly find billions of hard currency to build from scratch complex, maintenance heavy infrastructures in order to lift themselves into the 21st century.

What we can expect people to spend though is a few cents or dollars here and there. We've seen this in mobile telephony: if it works and brings tangible advantages to the user, then users themselves are more than willing to invest in new technology. And this is the case not just in wealthy, developed countries, but very much also in the crucial developing parts of the world. So what is needed is a system that allows individual consumers themselves to adapt the existing infrastructure, at extremely low cost, but with immediate, genuine benefits to themselves and their environment.

Another significant aspect to networked, digital technology is that with it in place you can start thinking in terms of applications rather than in terms of apparatus. This is a step at least as potent as was our step from using tools to powered machinery. When that happened, when we started to invent machines that use concentrated energy such as that stored in fossil fuels, for example by burning coal and turning the heat into steam, we accelerated enormously everything we put this kind of technology to, be they looms to weave fabrics or locomotives to pull trains. The same principle applies to petrol-fuelled engines or any other mechanical apparatus, even those powered by electricity. They make things go faster, they outperform human beings on many a scale, and there is absolutely no doubt that they constitute a milestone in our development.

But they too have their limitations, and these limitations are what we are approaching just about now. Because when you make a machine – be that a power-consuming or a power-generating machine – according to mechanical principles, then you have to first of all decide what this machine is going to be for: what it's going to do and how it's going to do it. Then you build the machine so it fulfils these criteria to the best of your intentions, and that then is what the machine is going to be used for, for as long as it's being put to its primary purpose. Once that's done, the principal variations you get are of levels or quantity. As you evolve the machine, it will probably get faster, you may add more functions to it, it may work with greater precision and become quieter, more reliable and use less energy, but the machine will always do what it's been intended for: churn a wheel, drive a turbine, spin a yarn, cut sheet metal, wrap boxes or do whatever it may be that you've opted for at the outset. Once you've refined your machine to a certain level, you will by necessity reach the outside boundary of what your machine can ever achieve. A car will always move from A to B. There are any number of things you can do to make that drive as smooth and as safe and as comfortable as possible, but its potential is really determined and set, and it is in essence just what a Mr Daimler or a Mr Royce had in mind for it right from the start, more than a hundred years ago.

With information technology, this is quite different: a computer does what the software you run on it tells it to do, which means that by changing the software, the application, you can completely change the purpose of your machine in an instant. The machine now becomes secondary. The primary focus is the application. If you use a computer you will see this borne out by your own experience. Even if you happen to be a great fan of a particular hardware product or brand, for example for reasons of design, what actually defines your user experience is the software. And here, there are no limitations to the potential, at least none that we can as yet discern. Which is why here we can start talking of *'potentiality'*.

This difference goes right to the heart of the step-change we're talking about. And it illustrates why we are talking about a different plateau or level that we are moving onto. Because up until the point at which we started making use of virtual technology, we were dealing with materials and mechanics and with *machinery* that is based on materials and mechanics. Of course there is potential in this, as we have seen and been able to exploit to tremendous effect. But that's just the point: we have seen and been able to use *to its limits* the potential of material and mechanics. A log of wood has the potential to provide a certain amount of heat when it's being burnt. A car with a 50 litre petrol tank has the potential to go a certain distance, at a certain speed. Or rather at a certain range of possible speeds. The construction of the engine, its parts, the materials used, will determine how fast, at the outermost extreme, it will go, how fast it will go safely and how fast it will go economically.

Once you remove yourself from the material world of potentials though, and step into the immaterial world of virtual potentiality, what you, or the thing you have invented, can do is no longer dependent on the material or the machinery, it's dependent largely on the software that you run. And software is programming: it's thinking. Intelligence, written down in code. The machine becomes almost – not quite! – immaterial. Of course, it still determines to some extent things such as processor speed and data storage capacity; but here, too, there is no limitation in sight: the computer that this sentence is being written on happily fits into a normal small backpack. It's currently attached to seven hard drives with a combined capacity of 12 terabytes. They take up about as much space as a nicely presented hardback edition of all seven Harry Potter books. But really there is no limit to what they, or the data on them can do: their capacity can easily be doubled, trebled, quadrupled, several times, and the machine that's attached to them will still be able to perform any task that any team of software designers develops for it. By the same token though, the moment the computer or its software break down, it doesn't matter how much machinery I have at my fingertips, it will be entirely useless.

The significant thing, therefore, the thing that *really* matters, is the application. So what counts is no longer quantity, the potential itself; it is, you could say, the *quality* of what we want to do with the potential we have. Potentiality.

Apply this principle to energy and you have a totally fascinating and in actual fact quite startling situation: the important thing, the thing that matters, is no longer the amount of energy, the quantity; the important thing is what we do with the energy: the *quality*. So when once what mattered about turning on a light was exactly that, for it to be light as opposed to dark, and when once the amount of light we got was directly related to the size of the candle or the strength of the turbine that powered the light bulb, now we are moving into a realm where the important thing is not the amount of light but the *quality* of the light we want. Is it for reading? For relaxing? Does it go with the music? With the furniture? How does it look in relation to the light that's coming in from the outside as the evening falls? In what ways does it differ when I watch a film from when I read? Or when I have dinner? When I have dinner with a date?...

We have arrived at a position where the point is no longer that the light is either on or off, or where we simply arrange lamps so they look nice and feel atmospheric. We now *orchestrate* qualities, applications and uses. And we haven't been here before: this is new. Which is why some of it sounds unlikely and maybe even borderline weird. It's a case of getting used to thinking about energy as something that is there in abundance and that can be arranged, used at will. Anything that someone may be inspired to develop a software application for, either today or at some point in the future, can be done. And that is also the reason why this is so important in the context of energy: it means we can't predict what uses a networked power structure can and may be put to. There are an infinite number of possible applications that have not yet been thought of, let alone developed; in other words: all kinds of potential are completely open...

IX   THE 'RESOURCE' THAT MULTIPLIES FOREVER

The analogy with smartphones is one we have invoked before and we will be doing so again, because it exemplifies potentiality so well: what that slim, smooth thing in your pocket can do is astounding not mainly because of what it's capable of as a little apparatus (though that too), but because of the potentiality it embodies. You can do with it whatever some software developer somewhere may think of.

And this entails one more crucial thought: when we develop applications, we don't use up any potential, any resources. Your smartphone does not become heavier, nor does it draw on any forests or mines of coal or tanks of oil because it can handle two million different apps.[XXXVI] This is only possible in the digital, virtual world. Without applications, you'd have to manufacture a calculator, a compass, a notebook, a telephone, an alarm clock, an egg timer, a dictionary, a camera, film for the camera, a photo album, paper for the photos, an address book, an encyclopaedia, a music player, a chess board, playing cards, a map, a radio and a television. And that's just to cover the basics a smartphone does before you add anything to it...

So there's a principal difference. Which is why you'll hear us and other people talk about a 'fundamental shift' (or a 'new paradigm') and sounding at times well nigh hyperbolic. Because with virtual technology it is not about using up resources and exploiting potentials, it's about the ability to create potentiality. And this ability *increases* the more we do it. This, too, is the opposite of what happens when we use stuff. When we use up material, the more we use the less we have. By contrast, when we use our intelligence and our ability to create and to learn and impart skill, then the more we do so, the more we have. Here, potential breeds potential. It was the American writer John Steinbeck [XXXVII] who said: "Ideas are like rabbits. You get a couple and learn how to handle them, and pretty soon you have a dozen." [XXXVIII] So when our consumption of goods and materials and resources leads us to diminish what we have and therefore generates a need to get more, resulting in a typical cycle of dependency, our use of ideas, of invention and thinking leads us to increase our understanding and our ability, thus generating yet greater capability to improve things further.

X   THE LEVELLER: DEMOCRACY IN THE NETWORK

Through this technology, then, our relationship with the technology itself and with what we use it for changes, be that information or energy. But what also changes, and in a way that is no less important, is our relationship with each other: the role we play in the constellation. This kind of technology creates a new relationship between users, producers and traders. Specifically, it takes energy out of the hands of the few who own the resources, and puts it into the hands of the many who have access to the network. This is rather different to where we've come from, and it's not necessarily a paradise, we're aware of that. The internet is a gigantic model for this kind of relationship, and it's the most impressive and most successful network there currently is, but it hasn't done away with concentrations of power: Google, Microsoft and Apple are not, as it happens, paragons of democracy and upholders of a level playing field, in fact the opposite; we have new behemoths with new ambitions for dominance and control. But here is the big difference to the old structure: there is no control in, or of, the network. There are hierarchies, true, and there are big players. There is commerce, of course, and there are some abuses. But there is no central control and there is no ownership of the network itself.

There are completely new types of behaviour, though, there are consumer-producers of content, and sharers, there are entirely new peer-reviewed and community-supervised barter, trade and loan schemes, and there is, without a shadow of a doubt, abundance: an abundance of information. And we are beginning to get to grips with it. Not quite without hiccups – that would be an unreasonable thing to expect, at any rate, since it's so new and we're so inexperienced at it by comparison – but we're learning how to handle abundance in the sphere of information. We can therefore assume, reasonably, that we'll get to grips with energy too, once we're dealing with it, and trading it, and distributing it, much along similar lines to how we now deal with information, in the network.

So do we actually have a cause? Is this some kind of revolution we are propagating?

Well, yes, we do have a cause. But it's not an ideological one, and it isn't borne out of environmentalism or capitalism or any other -ism that we care to think of. It's borne out of a strong sense that we have today in front of us a great opportunity that comes with great change and a great shift.

Energy is the lynchpin of civilisation. Without it, nothing goes. No person, no business and no infrastructure can function without energy. We may accept this as a given, and this is not what's going to change. But up until now we had to organise ourselves in this situation from a position where energy was *scarce*. Why was energy scarce? Because it was tied to resources. And we didn't even fare badly with it. We've achieved magnificent things, we've taken these resources and really made them work for us. We do not need to feel dejected about what we've done under the 'resources paradigm'.

But it has run its course. We know that it has, because on the one hand the resources are running low, and on the other hand, probably much more important, these resources, and our thinking that's tied into them, and our understanding of energy as something to do with them, far from opening up potential and possibilities for us any more – and they have done so in the past! – are now beginning to restrict us. We don't look at a petrol engine car and think 'what amazing things this could do', we worry. But worry is not a way forward. Worry is an ever growing obstacle to going forward. Yet we don't need to worry, because we have more than enough energy. We can largely forget resources and, to use the word just once more, emancipate ourselves from them. This opens up a plethora of new possibilities and potentials: potentialities. So our job is not to be scared of them or worry about them, our job is to cultivate them, to develop them, utilise them, invent with them. That means cultivate our ability: learn, practice, play.

Of course, we have no idea how that's going to pan out. What we do know is that everything to do with sustainability, efficiency and supply chains, all of which are of extreme importance as long as energy is a matter of resources, can be rethought and thrown open in new and never before imagined ways. And it also means that when it comes to 'progress' versus 'resistance to progress', to 'conservation' versus 'development' and to technophilia versus technophobia, our terms of reference are inadequate and outdated.

That's really all we know for certain: our thinking needs to move on. Everything else is more or less unknown. And that, in turn, is not new: we are, paradoxical as this may sound, familiar with the open and the unknown: we've thrived on it for millennia. It's what's very much brought out the best in us: curiosity, experimentation, innovation.

And that is why we say: relax! Take a deep breath and think of how fantastical a world you can create by taking a step forward. We can live in a world of plenty indeed. And we have the one 'resource' we need to do so, our intellect, our abilities, indeed our genius.

ENERGY IS NOT THE PROBLEM

III

Energy really, genuinely is not the problem. There are many ways to prove this, one of which requires taking out a calculator and going through a few computations involving pretty huge and therefore at first largely meaningless numbers. But even if the numbers are really too large to make sense of at a glance, it's nevertheless a useful exercise, because it will root our first and most fundamental premise in what we like best when it comes to 'proof': facts and figures. The premise is that energy – the quantity of energy at our disposal – is not the problem.

We start with the 'solar constant'.[XXXIX] Oft-quoted in relevant literature and on many a blog, this is the value of 1,368 watts (you will find this vary from about 1,360 to 1,370 depending on which source you follow), which is the approximate and fairly much agreed-upon amount of energy every square metre of planet earth receives from the sun. It's equivalent to approximately 126 watts per square foot, and it's almost the only 'constant' there is in this context.

Because from here on in, it becomes somewhat arbitrary: the value includes all types of solar radiation (what we experience as heat and light, but also invisible rays, such as infrared and ultraviolet light), and not all these types are easily convertible into energy we can make use of. Also, depending on what the weather is doing in any given place at any given time, more or less of it may get blocked out by clouds or absorbed by the atmosphere, or indeed, on a clear day, reflected back into space. Of course, energy that is absorbed by the atmosphere isn't lost: it becomes part of the weather, and some of that weather can again be tapped into, for example by strategically positioning wind turbines. Even so, letting that part of the equation go for the time-being, we still have around 700 watts of solar energy per square metre. Multiply that by the number of square metres of earth exposed to sunshine at any given time (about 50 % of the world's total of 510,072,000,000,000, so about 255,000,000,000,000, or 255 million square kilometres) and you get the difficult-to-grasp figure of 178,500,000,000,000,000 watts, which we can more manageably write down as 178,500 terawatts.

World energy demand in 2014 (the latest figures available as we revise this in 2016) was 13,699 Mtoe (million tonne oil equivalent), which is just under 159,319 terawatt hours.[XL] This is the energy consumption over a year. If you divide this number by the number of hours there are in a year, 8760, you get a figure of 18.18 terawatts: the global energy demand as power needed at any given time at today's consumption levels. It's no easier to imagine, this, than 178,500 terawatt, but the interesting point, and one that you *can* easily identify for yourself, is that nearly 180,000 terawatts is quite a bit larger than just over 18 terawatts, in fact, it is nearly *ten thousand times* larger. And that's why you'll keep reading, in our book as elsewhere, that 'we have 10,000 times as much energy as we need'. What it means is that the amount of energy *theoretically* available to us – nobody is suggesting in earnest that we could either now or in the foreseeable future capture *all* of this energy – is ten thousand times greater than the amount of energy we currently use.

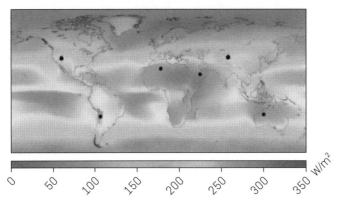

0   50   100   150   200   250   300   350   W/m²

The map shows average solar irradiance over three years from 1991 to 1993 (24 hours a day) taking into account cloud cover.

The areas marked out by dark dots could provide more than the world's total primary energy demand (assuming a low conversion efficiency of 8 %). In other words: all energy currently consumed, including heat, electricity and fossil fuels could be produced in the form of electricity by solar cells covering those areas.[XLI]

While we can't harness 100 % of solar energy that strokes our planet, it's really obvious that we have a whole lot of energy here that we are not using. As we've just seen, if we could get at just one ten thousandth of this energy (that's one percent of a percent or 0.01 %), then we would have our energy requirement of today covered. This in turn means that if we were able to make use of 0.1 %, or one thousandth part of solar energy available, we would have our energy requirement of today covered ten times over.

Now, it is entirely possible that we may, over the next 25 years or so, increase our demand for energy perhaps not tenfold but by a factor of two or three, even four or five. But these are not projections. The US Energy Information Administration (EIA) projects world primary energy consumption to increase by 48 % between 2012 and 2040.[XLII] So even if their calculation turns out to have been wildly inaccurate, and demand grows at two or three times the rate they expect, that would still mean that in order to cover *all* our energy needs by 2040 from solar, we would have to harness signally less than half of a tenth of one percent of the solar energy there is.

It's a tall order: in 2014, all types of solar, wind and geothermal energy made up 1.4 % of global primary energy production.[XLIII] That was up from 0.9 % in 2010, but we would still have to increase our energy output from these sources by about seventy-five times just to get to today's level of energy consumption.

Sticking with the US EIA forecast for the time-being and assuming that we need to cater for less than a doubling of our energy requirement by 2040, we can look at a rough target figure of some 30 terawatt. Bearing in mind that today's solar technology has a conversion efficiency of around 10 %, we would have to cover 0.5 % of the entire earth's surface with photovoltaic solar foils and panels if we wanted to cover *all* our energy needs from photovoltaics alone by 2040. This is not a fixed percentage, as it happens, but a conservative estimate. Already, today, there are photovoltaic cells that can convert up to 20 % of solar power into electricity, and there is nothing in science that says that this can't, or won't, increase further. But for the moment, why don't we err on the side of caution and assume a relatively low efficiency level, of the kind that you get with cheaper solar foils today.

So at that level, we would need to cover half a percent of the planet's surface, which sounds like an awful lot. Until you take a look at what we tend to use our land surface for now. For instance, there are in the world comfortably over 30 million kilometres of public roads.[XLIV] If you assumed an average of 15 metres width for each road (they do, of course, vary greatly) and covered them with photovoltaic panels, you'd already have enough to power the world. It's 450,000 square kilometres, which is just shy of the land surface area of Spain or about two thirds the size of Texas, or one third that of the Gobi desert. It's also about 1 % of the total land surface currently used for food production (49 million square miles as at 2011).[XLV] We'll be looking at some of these proportions in a bit more detail a little later, but for the moment suffice it to say that half a percent of the earth's surface can, if necessary, be found. We find it for all manner of things, we could, if we really wanted to, find it for the purpose of garnering completely clean energy.

But this is not in actual fact what we are proposing, nor, as far as we know, is anyone whom we take seriously. It isn't necessary, nor would it make sense, to cover *all* our energy needs by means of one technology, namely photovoltaics, even if we can assume, as we can, that this technology will become quite a bit more efficient than it is today, and a whole lot cheaper. There clearly are other technologies which, when used in combination with each other, will yield the kinds of results we are after much more effectively.

What we *are* proposing, as you already know, is that we shift our *approach* to energy provision, in its entirety. Because as long as we are focused on resources, the immediate, so as not to say 'knee-jerk', reaction not just to the little numbers game above (and we're aware that that's all it is), but to the whole idea of solar, weather and geothermal based forms of energy, is to point out their inherent flaws. For example:

- Two thirds of the earth is covered by water, not land, so anyone claiming that we'd 'only' have to cover 0.5 % of the earth's surface to get all the energy we need has already got that wrong: we'd actually have to cover *three times* as much, 1.5 %, of the available *land* surface.
- Even achieving a fraction of this would be an improbably vast undertaking that would have a massive impact on the look and feel of our cities and our countryside.
- Solar foils and photovoltaic panels use *a lot* of energy to produce, and they too use resources in their manufacture, some of them very rare.
- Many 'alternative' forms of energy, such as wind turbines, are unsightly, noisy, or will have some other impact on the environment, often substantial: hydrodams, for example, drown entire valleys to produce not much more energy than a small coal power station.
- And what if the sun doesn't shine or the wind doesn't blow: we'll need batteries for storage, which makes energy provision

expensive, and batteries too use a lot of resources in their manufacture.
• It's all of it expensive, uncertain, untried and untested, and we don't have much time for experiments.

Some of which may be true, some of which patently isn't, some of which is certainly debatable, none of which though is, for our actual premise, relevant.

We've already seen that energy is not the problem, there clearly is plenty of it. But neither is our problem what one person or another may like or dislike about one installation or another or about one type of energy generation or another type of energy garnering. In fact, none of these – often very justified – reservations and objections are our *problem.* They are practical hurdles. And as such they deserve to be taken seriously, of course, but that's about all they deserve. Being taken seriously, and handled as obstacles on the way.

I   DEALING WITH OBSTACLES ON THE WAY

Whenever you propose to implement, or are in fact implementing, a change of any description, you will come up against a raft of specific issues that have to be dealt with. The thing to remember is that that's what they are: issues to be dealt with, on a case-by-case basis, some of them local, some regional, some national, and some also global. They are the nuts and bolts of reconciling interests, they are the obstacles on the way. Every way, no matter which one you choose, will have obstacles in it, there's no point acting surprised about this, or throwing in the towel at your first glimpse of any of them. They may be what some people have a problem with, what they are *not* is the underlying, principal problem.

It doesn't matter what you are talking about, there will always be somebody who has *a problem* with it. And they may well have a point, who knows? Who's to say whose argument is valid in any given set of circumstances? There will be negotiations and arbitration in every place of the world where people with a mind and a voice to express that mind share a space. That's not going to change when it comes to energy.

You live in a beautiful mediaeval town full of heritage buildings and your neighbour wants to cover his terracotta tiles with solar panels. There's a tremendous view from your holiday cottage but since a few years ago, where there was nothing, they've now planted a humming wind farm. The artificial lake where you take your family for alpine rambling and that you think is in itself really quite beautiful could be raised by a meter or two, increasing the hydrodam's capacity by a good 20%-30%, but that will endanger the habitat of the marmot and destroy many rare plants. Alaskan oil fields. Open coal mining. Nuclear power stations. Methane gas tanks. Pipelines. Pylons. Fracking... And none of this, of course, is restricted to energy. Roads. Supermarkets. New housing. High speed rail links. Underground metropolitan transport. Entertainment licences. *Cycle lanes!* You change one bit of a part of something some people know, and you *will* get someone who'll object to it.

If there is a patch of grass near where you live, you can carry out an entertaining little experiment: no matter what state it is in right now, go to your local council and propose that you change it. We guarantee you there'll be someone who'll say 'no!' to you. If today it is lush and natural and full of yellow flowers, and you propose to cut it so the children can play football on it, the children's parents may be delighted, but someone will hate to see all those beautiful flowers go and worry about the bees having nowhere to buzz for their nectar. If today it's cut like an English bowling green and has 'Don't Walk on The Grass' signs all around it, and you propose that we could just let it grow wild, so that beautiful yellow flowers may blossom and the bees will have somewhere to buzz for their nectar, someone will hate you for ruining their pristine view from the balcony and making it impossible for them to sit out there any longer because of all the pollen wafting about that brings on their hay fever. If today it is an informal field that people lie and picnic on and you suggest how swell it would be if you had marked out tennis courts... – we could go on like this forever and the point would not become any clearer. *Anything* you do that changes an existing environment, even by the tiniest nudge, will get somebody's back up. So you can't consider *that* to be your insurmountable hurdle. That's just something you have to deal with and find an appropriate, fair and acceptable-enough-to-an-acceptable-enough-number-of people solution for. You want to change something, someone will be thrilled, someone will be upset, someone won't care. That's not your big problem, that's just the way it is.

And so you can instead concentrate on your *real* challenge, which is something completely different and has nothing to do with the 'amount of energy' there is or with the many and multifarious issues that crop up the moment you change any given set of circumstances anywhere in the world and that you invariably will have to address on a case-by-case basis.

Your real challenge is one of getting at the energy and taking it where it's needed, when it is needed, in a constant, dependable supply. Cheaply, efficiently, cleanly.

II   PREMISE PART I: THE ISSUE IS LOGISTICS

If your real challenge is getting the energy from where it is to where it is needed, then you have an issue *not* of supply, *not* of environmental changes or local and regional obstacles, you have one, very plainly, of *logistics*.

This is the first part of our premise. There will, as you might expect, be three, but number one is: *energy itself is not the issue, the issue is logistics.*

The *reason*, we have seen, why energy is not the issue is because we are not in fact dependent on resources any longer. It is resources and our dependency on them that has made energy scarce. But, you may point out, and rightly, we haven't actually *solved* this dependency yet. Far from it. You will be able to cite disheartening statistics that show just how little we have weaned ourselves off resources to date and how much we still burn coal and oil and gas and feed our nuclear power plants. We will get to this. For the moment, let us just state that we have extremely good reason to expect that we are about to see not just gradual, incremental changes to the proportion of our energy demand that we cover from infinite sources of energy, but *exponential* changes. And of course we'll explain why. But before that, let's deal with the second part of our premise.

III   **PREMISE PART 2: WE HAVE ENERGY THAT GETS CHEAPER**
        THE MORE OF IT WE USE

Much of this abundant, infinite source of energy – the sun – is accessible to us directly in the form of electricity, by means of solar photovoltaics. Again, we'll explain how they work and why this is the case, so if you have grave doubts then park these for the moment, as we look at one particular aspect that is *unique* to photovoltaic solar and that changes everything we used to know about energy.

Up until now, because our energy has been coming from resources, which are or will become, at one point, scarce, energy has followed the standard laws of supply and demand, whereby the more people use it the more they need it, and the more they need it the more they compete for it; and at the same time the people who provide it keep improving their efficiencies and their methods and so there is a continuous, but fairly linear tug-o-war between those who have it and those who want it, meaning that with some exceptional fluctuations, energy prices rise steadily (and, as it happens, over the long term, quite slowly). This may sound counter-intuitive, as, especially at times of 'crisis', we may experience sharp hikes in petrol or electricity prices. But in actual fact the long term trends for coal, oil and gas are all on a slow steady upward inclined slope that is surprisingly even, a bit as below. (Note that this is deliberately simplified: the actual curve shows spikes in coal prices during the war years for both world wars and a significant one for oil during the 1970s oil shortage, but they level out again fairly smoothly as soon as each period of distress is over.)

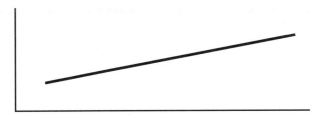

Now with solar – and this really applies very specifically to photovoltaics – we are putting energy on a completely different trajectory. And we say it again because it's important: this hasn't happened before in our history. For the first time ever, we can use energy *not* by going through mechanical processes or some natural energy stores, but directly, without detour, from the sun. We are very simply, and very categorically, putting up a piece of semiconductor material – a solar foil – and it garners energy. Energy that is immediately usable and that can be sent anywhere, even hundreds, indeed thousands of miles away, in an instant. No pipelines, no ships, no lorries have to take any material anywhere. You can connect the foil or panel by a wire to anything you like – a machine, a light, a power grid – and it will be able to feed energy to it. What this means, and what is so interesting about it, is that we are not 'using up' the energy, we're just redistributing it. That's all. And the thing we're using to do so is a bit of foil or panel. What we have to *make*, or manufacture – produce in really large quantities – are these foils and panels. And these foils and panels are semiconductors. They're the same type of material that you use to make computer chips. They're made of silicon, which is sand, and a few other materials. Again, we'll come to the specifics in a bit.

And this in turn, as evidenced in long term global data, means that their manufacture is subject to *Moore's Law,* they become cheaper and more efficient, *the more of them you make.* They are on an exponential downward curve, a bit like this:

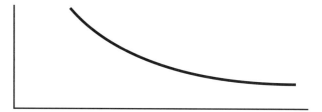

And this means that energy, for the first time, is on the same trajectory as *information.* Which is why we ventured a little earlier that we find ourselves in the extraordinary situation of being able to say: 'kilowatts as kilobytes – the more we use, the cheaper they become.'
Overlap the two shapes like this:

And you see that there will come a point at which the two lines intersect and energy from the source that started off very dear is now cheaper than that from the other sources, and continues to be so, no matter how much of it we use.

These stylised drawings may do little to convince you, but they are in fact born out exactly by figures. In his guest blog on Scientific American, Computer Scientist and entrepreneur Ramez Naam states: "The cost of solar, in the average location in the US, will cross the current average retail electricity price of 12 cents per kilowatt hour in around 2020, or 9 years from now. In fact, given that retail electricity prices are currently rising by a few percent per year, prices will probably cross earlier, around 2018 for the country as a whole, and as early as 2015 for the sunniest parts of America." And he illustrates this with a graph very similar to the purely abstract ones we've drawn: [XLVI]

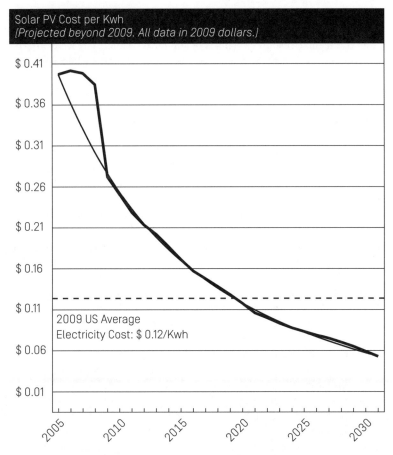

Source Data to 2009: DOE NREL Solar Technologies Market Report, Jan 2010; Projections by Naam 2011

That moment, the moment when energy from photovoltaic solar becomes cheaper to the consumer than energy from traditional fuels, is of significance to everyone no matter what part of the energy sector they're in, because cost is a massive driver of demand, as you know yourself from experience. When solar is cheaper than coal, oil or nuclear, people will *want* it. And they'll be able to get it, because it's not going to run out. There may, as has already happened once or twice, be short term blips while solar panel manufacturers won't be able to keep up with demand in real time. Let that not distract you: if you want to know what will happen to the cost of solar foils, look at what happened to the cost of USB sticks. They started out at $60, $80 a piece for a tiny amount of storage, maybe 32 megabytes, and now companies big and small will happily give you a USB stick with several gigabytes capacity for free for your keyring, as a marketing gift.

There has been talk for a while now of 'grid parity': a situation when energy from 'renewables' costs the same as, or less than, energy from 'traditional' resources. This is of similar relevance, but we think the most dramatic and most 'revolutionary' moment will come when, very specifically, electricity from photovoltaics becomes cheapest. That will be the point at which the balance will tip. Nobody knows with any certainty *when* this moment will come, but it will come. You could call it 'e-Day': the day that everything to do with electricity will be turned inside out, for good.

So we have energy that becomes cheaper the more of it we use. And what goes hand in hand with it, and is no less significant, is this: with resources, the more of them you use, the less usable energy you have left. It's the substance-based principle: if you have a shed full of wood, then the more of the wood you burn, the less energy will be left in the shed for you to use. If you want more, you'll have to go and get some. The same applies to oil under the seabed or coal in the mine. The more of it you use, the less you have available. All our most important resources, coal, oil, gas, uranium, are finite. There is a stock and when it's used up it's gone.

How big the stock is, and how long it will last, that's open to a great deal of debate, which we don't need to get into now, because there are any number of books and websites that will tell you in just as much detail as you care to know. What is of real interest is the fact that with photovoltaics the direct opposite is the case: the more foils or panels you produce, the *more* energy you'll have in the long term.

Instead of resources being *used up,* materials are being *used* to make available *more* of the energy, continuously, and into the future. It's a bit like planting saplings or seeds: the more of them you invest in, the more they will yield. So you are, in a sense, applying a farming principle to energy garnering, except you don't have to go through the stage where the energy first turns into plants or animal feed or food and then gets turned back into energy again. You have it available to you as electricity straight away.

And so if the first aspect of our premise is that energy is not the issue, but logistics are, then aspect number two of our premise is this: we are talking about abundant energy that not only becomes *cheaper* as we use more, it also gets *more* as we use more.

IV   PREMISE PART 3: THE KEY IS NETWORKED ELECTRICITY

The third aspect to our premise follows on from the first two: if we are going to avail ourselves of this abundant source of energy and treat not the amount of energy as the problem, but the logistics of getting it to where it's needed, and if, in tandem with this, we are going to have energy that becomes *cheaper* the more of it we use, while at the same time making available *more* energy, the *more* we use, then there is only one principal 'mode' of energy that we can be talking about, and that is electricity.

Electricity is the only way that energy can be 'medial', meaning that it can carry information about itself at the same time as carrying itself. We may, as a society on the whole, not yet be quite aware of just how brilliant this is. We tend to look at electricity itself simply as another resource. We pay gas bills and electricity bills. We choose between petrol-fuelled cars and a growing selection of viable electric cars. We can either use oil to heat our houses, or electricity. So no wonder we categorise electricity as the same 'kind of thing' as gas, oil or petrol. *But it is nothing of the sort.* Electricity is – and this is a point we'll be elaborating on a fair bit later on – in a sense an *abstraction* of energy, meaning that it is not the energy itself, it is the energy turned into something else: a meta-form of energy, so to speak. It hurts the brain a bit to think this through, but it's worth making the effort: gas, oil, petrol, coal, wood, biofuels, solar fuels: they're all just energy stores. They are nothing more and nothing less than substances which are capable of containing and therefore storing energy in such a form that we can get at it. The *energy* itself is not the gas, the oil, the petrol, the coal... the energy sits in the substance. We, by burning the substance, release the energy. And one of the most practical and at the same time 'magical' things we've ever learnt to do is to then turn that energy into electricity: a form of energy once removed from the substance. The same applies even when we garner solar power and turn it into electricity. Now we don't deal with substances as energy stores any more, but electricity is still a converted form of the energy we started out with, in this case light.

As electricity, energy travels at very nearly the speed of light and it can do anything we want it to do: power machines, heat or cool buildings, drive cars. *Beyond* that, it can also deliver information. And it is this combination that makes electricity so tremendously useful and versatile. Because it means that unlike any other form of energy it can be *networked.*

'Aha,' you may say, 'what nonsense! Other forms of energy can also be networked: take the really quite remarkably impressive remote central heating system in use in Basel, Switzerland, where energy as heat is produced in a large waste burner at the edge of the city and then channelled through a system of pipes round many people's houses. That, surely is a network.' Yes it is, and no it isn't.

The term 'network' is one of those we've come to rely on perhaps a touch nonchalantly because it so efficiently serves as a shorthand for a *particular type of* network. Of course, there are in the world all manner of networks – gas networks, water networks, traditional electricity networks – and they work perfectly well as what they are: top-down hierarchical distribution systems. The word 'network' describes these very inaccurately though, because if you look at how they're structured, they are not the way you would tie a net, they tend to start with few nodes at the top (the power station, the water works, the gas works) and then branch out to more and more nodes until you have the individual power sockets, gas fires and water taps in people's houses at the very end. But try to send some of the gas that's coming out of your cooker back to the gas works or across to your nephew out in Metzerlen, and you're in serious trouble.

The way we mostly use the term 'network' today is much more useful and accurate, and it is also what we now mean and refer to in this book: it's a network which can grow parts of itself 'organically', simply by connecting more nodes big or small to existing ones, and in which there may still be hierarchies (we've touched on this briefly once before), but the flow of whatever is being exchanged is multi-directional. It's the *rhizome* concept: a pattern or a shape that has points and nodes connected with each other in all directions in non-prescribed, potentially ever expanding and ever-intensifying structures and sub-structures.[XLVII]
These three elements together, then, make up our premise:
- Energy is not the problem, we have more than enough of it. The challenge is logistics.
- With photovoltaics in particular, the principles of energy and information technology are coming together, and energy for the first time is now subject to Moore's Law: it's becoming cheaper, the more you use it. And, again as in computing, the more hardware you devote to it, the greater its potentiality. The material you use for making the hardware is not burnt or used up, it's there to make more energy available to you. And this applies not just to photovoltaics, but to any of the infinite sources of energy we have, such as wind or tidal, for example.
- The only way that we can address the challenge of energy logistics and make use of energy that is subject to Moore's Law is by dealing with energy primarily as electricity. Electricity is what makes the coming together of energy and information technology possible, because it allows us to network energy in a meaningful way.

You may notice that these three elements are not all the same. Point 1 is a statement of fact, as much as anything can be said to be 'fact'. Point 2 is an observation on a current technological development. We cannot say with absolute certainty that this is so, we can only say that this is what we see happening, and because to us it is plausible, we can incorporate it into our premise. Point 3 then builds on the observation we've made in Point 2 and says, if that is so, then what we may infer from it as an appropriate way forward is this.

V   CAN DO VS HAVE TO

One thing that has struck us over the years that we've been looking into the issue of energy and formulating our thoughts on it is how much we've been hearing and reading about what we *have to do*. We *have to* reduce carbon emissions, we *have to* preserve our resources, we have to limit our growth... Very little was being said about what we *can do*.

But there is a big difference between being forced, or coerced, into doing something, either because there is no alternative or because there appears to be none, and *being able* to do something, because it makes sense. Or because we feel like it. Or because it's good for us. Or because it's fun. Or simply because we can. Ability is a much greater power than mere force, as anyone with children will know: if I tell you to do your homework because you *must,* then I may just succeed in getting you to sit down and do your homework, but in all likelihood that's about it. If on the other hand I enable you to do a project because I inspire you and give you the tools to do it, chances are you will take this somewhere else completely. It may sound a bit whimsical to put this in such simple terms, but there is a serious and overriding thought behind it, which is this: there are two possible stances we can take. One is to say, here is a grave situation, we must deal with it as a matter of urgency, and in order to deal with it we must refer to and draw on everything we know, because there are certain ways to do things and the matter is serious enough to warrant that we do this 'properly'. The other is to say, here is an interesting, though challenging, situation. It's certainly serious, but it also opens up all manner of possibilities, and we may be able to deal with it in ways we've never even imagined before: let's see what we can come up with.

The reason we advocate the latter is not arbitrary. It's not just that we like the idea of it and think this would be more enjoyable and easier going. It's also that we are precisely at the juncture in our development where we actually *can* take this position, because we have technology that quite literally *enables* us.

Earlier, we talked about potential and potentiality. Let's explore this a little further. We ventured that in matter, in resources, there lies a certain quantifiable potential: a log of wood has the potential to give us a certain amount of heat, for example. This is applicable to all resources:

in the as yet unexploited Alaskan oil fields, there lies the potential for a considerable amount of energy to be released and used. This is the way we traditionally understand energy. It does entail potential, and it also entails ability, of course. Given a certain amount of energy, we can do certain things: heat homes, drive cars, fly planes, make products. So there is potential, and we understand how to release this potential. But as we have also seen, the potential, in being used, over time disappears, because the material that gives us the potential is being used up. What we are doing is *consume*. This permeates our very being: we think of ourselves, even readily identify ourselves, as 'consumers'.

Very understandably we therefore begin to sense unease and discomfort when we realise that the consumption, the using up, of resources leads to these resources running low on the one hand, and to the products and by-products of what we've turned the resources into damaging the world we live in, on the other. We now have floating about in the oceans gigantic man-made patches of rubbish, covering hundreds of thousands of square miles, consisting of plastics, chemical sludge and assorted litter.[XLVIII] The millions of tons of carbon dioxide that we release into the atmosphere every year in the process of using fossil fuels are subject to global conferences and coordinated emergency measures, and recycling has become a watchword for domestic as well as industrial waste management.

Very obviously, consuming things made of stuff results in that stuff ending up as something else, and the limits to what we can do are set by the planet's and our own capacity for a) providing the raw materials of which things are made and b) handling the waste that these materials turn into.

Now, we've also already mentioned the coming together of information technology and energy technology. And we've said that this opens things up in a new way, because in information technology, we are no longer just looking at potential, but at potentiality. Here, because we're in the realm of the virtual, our ability is not determined by material, but by the thinking that goes into how the material is put to use. With a log of wood, our potential for warmth goes as far as the energy contained in that log of wood, *and no further*. With a photovoltaic solar panel, our potential goes as far as any software application currently in existence for the use of the energy it harnesses as electricity makes possible, and beyond that to any software application that may ever be invented.

You could argue, of course, that the solar panel too has its limitations. You can point out that the amount of energy it can convert into electricity is determined by its size, by the materials that it is made of and by the amount of sunlight it receives. And you can furthermore point out that a log of wood converted into heat and that heat converted into electricity *also* yields access to the potentiality of applications. All that is true. But the relationship between energy and matter is absolutely different. In the log of wood, the relationship between matter and energy is linear and finite. X amount of wood will yield Y amount of energy, and then once you've burnt it, it's used up and done. With the solar panel, X amount of surface area will yield Y amount of energy, under a certain set of circumstances, over some time, and the material doesn't go away nor change. Over some extended time, such as 15 to 20 years, the panel will 'wear out' and not be able to convert light into electricity efficiently any more, but it will be easy then to recycle the panel and turn the materials in it – which haven't been used up, they have just been reconfigured slightly – back into another solar panel that can harness another 20 or 30 years' worth of energy.

We will get into electricity and what it actually is a bit further down, but for the moment hold on to just this thought: with photovoltaics, we are able to make use of an inexhaustible source of energy and turn this energy into electricity directly, without waste or by-product, and without any intermediate processes. And this is not energy that we 'consume' in the material sense we're used to: nothing gets 'used up'. We *are* using materials to make the foils or panels which convert the energy for us, but these materials stay intact. When before, the more material we used to generate energy, the more waste and by-products we produced, we can now use materials to harness energy without any by-products or waste at all, and when these materials come to the end of their useful 'life cycle', we can reuse them. This means we are going straight from abundant energy to potentiality, from the sun in the sky to anything we care to imagine.

And that is why we no longer want to hear about what we have to do, but want to talk instead about what we can do.

VI   FORMULATING A DIFFERENT KIND OF QUESTION

Some time between about 1.9 million and 400,000 years ago, early humans embarked on a tremendous adventure. They invited into their homes and into the very centre of their existence one of the most powerfully destructive forces nature has to offer: fire. The effect this had on our development was startling. Using combustion in a controlled way allowed our distant relatives to cook food, which meant their diet could become much more varied and nutritious than it had been, and with the same single fell swoop, much more of their time and energy could be spent doing things other than simply digesting. Plus it allowed them to keep warm in winter, to make the night visible, to scare off dangerous predators, and to proceed to crafting much stronger materials than they'd used before. It is no exaggeration to say that fire really put energy into our hands and that energy brought us civilisation, with everything that entails.

The impact fire had on us as a species can hardly be overstated and so, naturally, fire shaped our understanding of what energy is. And so it isn't really that surprising that for all the progress we've made, in many respects not much has changed since those heady but prehistoric days: we still use fire to cook and to keep warm. Fire still drives our development, fire is still what we use for energy. Granted, we now turn a lot of that energy into electricity, but the principle is pretty much the same as it has been since the very beginning: when we need energy, we for the most part take some substance, be it wood or coal or oil or gas, and burn it; some of it at home, much of it in power stations. That's the essence of it. At the very basis of what we would recognise as 'modern' living lies a concept of energy that hasn't shifted since shortly after we got on our hind legs and started to walk upright.

No wonder then, that we see energy as something that is or will become scarce, as something that's tied to 'stuff' that we need to burn. But while this concept of energy has got us a long way and set us apart from any other species on the planet, holding on to it is, as we have seen, neither sensible nor necessary any longer.

So from our premise, and with our perspective, we now realise that we can ask a different kind of question. We've stated earlier the pretty obvious fact that asking the right kind of question is precondition to obtaining anything that resembles a 'right' answer. This in itself is something we can formulate differently. Not in terms of 'right' and 'wrong' but simply in terms of 'corresponding'. When we talked about the 'energy question', we said that it is not 'have we got enough for everyone?' but 'how are we going to organise it?' This is still, however, a problem/solution line of enquiry. It's a more specific way of saying: 'what is the problem, and how can we solve it?' But we can already go one step further than that, and ask a different *category* of question. Such as: 'what is the vision?'

What is it we actually want to achieve? Not in light of what we perceive to be our current energy situation, not even in generally pragmatic, utilitarian terms, but *in principle.* What kind of world do we want to live in? If we work from an assumption of energy *abundance,* we don't have to ask ourselves constantly 'what is possible within the narrow parameters set by our limited resources and the limitations imposed on us by an ecosystem that suffers from our use of these resources?' We can ask ourselves instead: 'What is it that we, with our intelligence, our imagination, can and want to come up with? What can we make *better, more interesting, healthier?'* Because, and this is absolutely central to our argument, if it isn't resources that we are limited by, then what *are* we limited by, other than the scope of our own intelligence, our own imagination, our own – let us use the word – *genius?*

Since we are asking this question, we may also take this opportunity to clarify what we mean by 'genius'. This is important, because we don't use genius here as a self-congratulatory term to pat ourselves on the back as individuals. Genius we can use here in the sense of our collective intellectual and creative power: our joint ability to invent, develop and hone technological solutions which work together, across national and cultural boundaries, in multiple layers of networks that span the entirety of our species. Genius here is not intellectual brilliance or outstanding ability in one person, it is our capacity as a species to create and cultivate deliberate, specified qualities.

Once we start thinking in this way and begin to understand ourselves and our situation in this broader, in truth global, context, we open ourselves up to a different modus. You could say 'stance', but, as we've noted before, that implies standing still, and we will not be standing still. Anything but. We will be even more in flux, in motion, than we always have been. That in itself, this great flexibility, is something we're still getting used to. Since we are not going to be restricted by external limitations and restrained by centrally organised, controlling structures, we can behave with a great deal of possibly quite bewildering freedom.

We don't have to subscribe to sets of views, and we don't have to think up and carry through a master plan. We don't, in other words, have to make up a Utopia for ourselves and then attempt to reach it; something we would invariably fail to do, because one person's Utopia is another person's Hell.

Instead, we can pick and choose ideas and cultivate them to our wants, needs, desires and imaginations. You might call it the Flower Shop Principle: you walk into a flower shop and the friendly assistant comes up to you from behind the counter and asks you 'what would you like flowers for today?' And depending on what you are hoping to use your flowers for, you might say: 'oh, perhaps something a little subdued and subtle, it's for a dear friend who has passed away,' or you may say, 'something bright and colourful that will cheer me up in the morning when I come down to my breakfast table'. We are now pretty much in a position where we can ask ourselves, 'what sort of arrangements of ideas are we after today?' and we can answer, 'well, it would be good if we got rid of carcinogens in the city air,' for example. Or we could say, 'I need something exciting to do for my 50th birthday party.' In either case, the modus we are in is one of picking some ideas that may help us move towards the desired situation.

There is no need then, for us to claw back and say to ourselves: 'Ah, but that can't be done, and it's never going to happen.' Much as we wouldn't stare at the flower seller and say to him or her, 'what a stupid question, you silly oaf, don't you know that it's just flowers you're selling, how can I be after any meaning at all, it's just dead plants with some green stuff and colours on top,' we don't have to say to ourselves: 'don't be ridiculous, ideas are never going to get us anywhere.' Because the operative question in this equation is not, 'what is possible now with what we've got?' the operative question is, 'what is *potentially* possible, with everything that we have already invented and may yet invent?' And *then* we can look for practical and political ways of dealing with these ideas.

Once we have a vision of what we *ideally* want our world to be like, we can begin to build our pathway there. Step by step, with setbacks and obstacles along the way, but in the knowledge that there's somewhere worthwhile to get to at the destination. And there is no doubt that by the time we get to the destination, whether it be our actual goal or whether it be an approximation to it, new goals will come up on the horizon. This is not going to be the journey to end all journeys, this is no more, and no less, than another, but significant, leg on a very long trip indeed. And this in itself really isn't new at all. Throughout our entire history, from our very earliest ancestors onwards, we have been travelling like this. And there is no reason why we should stop now.

VII   THE VISION: 'ALWAYS ON' POWER

So here, then, is our vision: a world in which the power is always on, in which nobody 'owns' energy, in which energy is 'just there'. So much so that we can forget about it and concentrate instead on other, more interesting and enjoyable things.

The fact that this is possible now marks an extraordinary step in our development. It means we can leave behind a mode of dealing with energy that corresponds to the neolithic era, and effectively commence a new era, one that is based not on substances and resources, but on the next stage beyond that: data and intellect. And this is not a case of throwing out the old and bringing in the new: we have built our entire civilisation on a long and complex series of discoveries, achievements, setbacks, failures, new successes and amalgamations of experiences. We treasure these and know they are precious to us not only in a practical and applicable sense but also in an emotional sense: our values, our identity, our culture have all been defined, over a long time, by the age of stone and fire. We are now, though, in a position to carry this into a next chapter that takes us into the realm of potentiality and the virtual, *beyond* materials, substances and resources. One we feel inclined, therefore, to call the *metalithic* era.

VIII   OUR 135-YEAR LAG

When we talk of a 'new era', and go as far as giving it a name, we're aware that we may sound not just provocative but quasi evangelical. But there is nothing so new about it that we could stand here today and say: 'Look, people of the world, we have a whole alternative truth for you, harken up and be *amazed*.'

The concept of energy that we are talking about has been becoming familiar to us over the course of the last 135 years in the form of electricity. So, in actual fact, we're lagging behind ourselves. For about six generations now we've been making use, in ever more inventive ways, and with ever more spectacular success, of energy in a form that is completely *without* substance. Energy that isn't 'stuff'. You can't touch electricity; if you touch an electrically charged wire, you don't feel the substance of energy, you feel the effect of it. The energy itself is intangible.

And think of the extraordinary transformation that has happened to our lives within only the last few decades: on 4th May 2011, as we were writing this particular sentence, Claude Choules,[XLIX] the last surviving veteran of World War One, died in Salter Point, Australia, aged 110. On the day he was born in England, 3rd March 1901, Queen Victoria had just recently been laid to rest. There weren't as yet any trains or trams running on electricity, there wouldn't be any computers for a while. London and Paris, the big European cities of the day, were mainly gaslit and heated by coal, while in the new world, in New York, something interesting had started to happen...

Down in Lower Manhattan, not far from the Brooklyn Bridge and the World Trade Centre, at No 257 Pearl Street, stands a not very impressive, eighties-style, red brick building with an uninspiring retail floor that features a non-exceptional bank branch. It's entirely un-note-worthy. Or would be, if it weren't for the fact that on this site, on 4th September 1882, at 3 o'clock in the afternoon, history was made in a way that would change everything, for everyone, for good.

Here, Thomas Edison's Illuminating Company[L] had built the first commercial electrical power station in the world, starting out with one large dynamo that was capable of powering 400 lamps for 85 customers in the area. On that day, the power station went into operation, and the lights in Manhattan went on for the first time, powered not by gas, but by electricity. The technology quickly caught on, and two years later the company was serving more than five hundred customers with over ten thousand lamps.

Without a doubt, the reason for electricity's success was not just its novelty and the warm, steady glow that Edison's light bulbs provided, but in particular the simple fact that it was safer and, most crucially, cheaper than gas. Following high initial outlays for the purchase of the site and the installation of wiring and meters, and in spite of considerable maintenance costs, the enterprise became profitable after only two years. By the time the station burnt down in 1890, it had established itself as the model for most of the electricity supply systems that rapidly followed.

At his power plant, Edison burnt fuel to heat up water in large boilers which stood in a building adjacent to the one housing the dynamos, and then fed the steam produced by the hot water through to the dynamos to generate electricity. Since the Pearl Street power plant went into operation 135 years ago, electricity has become omnipresent in our lives and almost everything about how we experience life has been altered as a direct result of it. But in terms of electricity generation, not that much has really changed at all. And *that's* wherein lies our problem. We've created a peculiar disconnect between what we are doing and how we are thinking. Today, we are using electricity in ways that Queen Victoria would have considered wizardry. Yet somebody born barely a month after her funeral lived to see live pictures from the moon, MRI scans of the body and everything that's on the internet. In the Academy Award winning film *The King's Speech,*[LI] Queen Victoria's son, Edward VII, rues the day when radio enters his life and broadcasts his speeches to the country and an empire that spans the globe. His son, George V, then overcomes his speech impediment and turns radio to his and his nation's advantage. His grandson, Prince Charles, went through one of the most public, and publicly humiliating, marriage breakups of the 20th century with Princess Diana, and their son in turn, Prince William, got married to Catherine Middleton, on the 29th April 2011, in front of a TV audience of two billion people in 180 countries.[LII]

What does any of this signify? It signifies that it took no more than five generations for our understanding of ourselves, our ways of communicating ourselves to each other and our way of 'doing things' very generally to change beyond recognition. And yet while all this has been happening, we *still* think of energy not as something abstract, without substance, as something that is simply there, for us to capture and to manipulate, to send across a network as we please, but as something that we have to dig out of the ground and channel through pipelines across continents. Our perspective on energy is 135 years out of date.

IX   ABUNDANCE

We cannot start to think of energy as something we have in abundance as long as we see it as tied to substances and resources. The moment we recognise it as something abstract, something intangible, however, it immediately becomes obvious that of course there is plenty of it around. Just as there is vastly more than enough air around, so the earth basks in free energy.

Let's briefly recap: why do we have an energy problem today? Because we look at energy as 'stuff'. For 'stuff' – we realise it's an annoying catch-all – read 'resources': coal, oil, gas, the usual. This is ingrained in us; it's a mindset. And there's nothing wrong with mindsets, we need them to make sense of the world. But the world changes. Mainly it changes because we change. If you are uncomfortable with the idea that it is because of us human beings that the world changes and feel inclined to point out that without us, too, the world would be changing and that it *is* changing with us on it in many ways that we have no control over, then of course you are right. But in ways that are meaningful and relevant to us, the world changes mostly because of us. You may think that's human-centric, and it is. Because we are the species that, more than any other, is shaping this planet. And *that* is not something that's going to change any time soon, like it or not. What *can* change, what is entirely within our grasp, is the kind of impact we have on the planet.

We have come a long long way since the days when we first emerged from the forest and started to plant things, cultivate the land and farm animals. It is time for us to realise that energy isn't stuff: burning stuff is just one way of getting at energy, the energy itself comes from the sun. It gets absorbed by plants and trees and microorganisms that then die and turn into coal and oil, or if it's wood it gets chopped down and burnt before it has a chance to do so. But, with some notable exceptions, *all* energy that we have here ultimately comes to us from the sun. And there are much better ways of getting at it than by burning fossil fuels. And even for the fuels we need or want to keep, there are now ways of *making* them, directly with energy that comes from the sun.

But what does 'abundance' actually mean? Is it necessarily such a wonderful thing to have abundance? Perhaps it isn't, perhaps it is. We are not able to say, because it very much depends on what we do with it. There is nothing *inherently* 'good' or 'bad' about abundance, it's not something we can classify morally. What we do know is that it's categorically different to the opposite, which is scarcity.

And why should this matter? It matters because apart from the purely practical aspect, that which looks at whether there is enough energy to power a city, for example, or a country, there is also a fundamental, if you like philosophical, aspect to this. One that looks at how we do things *in principle.* This may not, at first glance, seem entirely relevant or to matter that much when we are facing so many hands-on, practical issues, but it does matter. Enormously. Because what are hands-on, practical issues other than the limitations that we face to becoming what we could become? It goes right to the heart of how we define ourselves as human beings.

X   THE FEAR OF GOD

One of the big news stories that emerged while we were writing this book was the announcement, on 4th July 2012, that scientists at the Large Hadron Collider at CERN [LIII] had found the Higgs Boson. [LIV] This was an extraordinary milestone in science, the significance and further ramifications of which will take some time to become fully known. There are many things that interest us about this, but none of these can we or do we want to delve into here, other than one tangential little fact, which, as it happens, rather irritated Peter Higgs himself: that his boson became popularly known as the 'God Particle'.

We are not going to go into any theological debates, nor are we going to get into religion. What we do want to briefly touch upon is the idea of where we are at in terms of knowledge and power. As a species, we've been in a position where we've been able to destroy ourselves and most advanced life forms on planet earth, for about sixty, seventy years now. We have made ourselves capable of an unimaginable destructive force, and we have had to rein in this capability. We've had to go against our previous instinct, which was to *use* every tool and every weapon at our fingertips, and instead say to ourselves: let's *not* use this force. What that has done to us is bring us closer together and address some of our conflicts and problems on a global basis, and so far – the many local and regional conflicts that still rage notwithstanding – it has worked.

At the same time as putting in place this incredible destructive power, we've also developed a magnificent constructive tool: the internet. Today, we are able to share all the knowledge we jointly have as a species with each other, in an instant. For free. This was not only unheard of, it was undreamt of, only a generation ago. We ourselves, the people writing this, remember vividly a time when we could simply not have imagined that this would be possible, that's how new, and how groundbreaking this particular achievement of humans on this planet is.

We can spin this thread a little further back and ask ourselves: what would people a few hundred years ago have made of the idea that there would be a time when we have universal education. Never even mind the internet: just a culture in which every boy and every girl goes to school for nine years at least, as a matter of course. By many a mediaeval monk or Renaissance cleric this may not have been looked upon with unalloyed joy. If you give knowledge to the masses, then you give them power, and if you give them power then you take that power away from the authority that 'knows what is best'. Knowledge is *dangerous.* Malala Yousafzai [LV] knows just how dangerous: on 9th October 2012, she was shot in the head by the Taliban in her native Pakistan for no 'offence' other than that she defied their edict that girls should not go to school...

But extremists aside, today, across a majority of countries around the world, we have universal education and the world has not ended, it has vastly improved for it. Now, with the internet, we've taken the concept of universal education another leap forward: now we have universal *sharing* of information. And we have this *abundance* of content. Are we sinking under the weight of it? Not really, we are learning to index, manage, handle it: we are learning, as the metaphor suggests, to surf on it, and the connotation of fun and enjoyment, of pleasure and random discovery that comes with the term here is entirely appropriate.

Could we have foretold what today was going to be like, fifty years ago, let alone five hundred? No. Should we, therefore, be deeply concerned if we can't foresee or foretell what a world will be like in which we also have an abundance of energy that we can share in the network? Not really, no. It's just not really possible for us to know. We will, invariably though, find out, and there is no reason to expect that we won't be able to make it work. Experience, for all the ills in the world that still persist, suggests that we will.

'God' tends to be described as the all-knowing and the all-powerful being, and as the unknown and unknowable. Which is why God needs to be both revered and feared. So what happens if *we ourselves,* the human species, become ever more knowing and more powerful? What happens if we put into our own hands the means to share everything we know and give ourselves all the power that we could ever want? Do we need to fear ourselves? Or do we have to find ways of knowing ourselves in commensurate ways?...

XI   MORE IS MORE

We're going to touch down on some very prosaic and practical considerations in just a moment, together with a whole lot of facts and figures.

But allow yourself this thought: what if we were to say to ourselves: 'Let's use more energy, not less!' – On the surface of what we've known and accepted as our reality to date, that would be insane. The received wisdom is that we need to restrict ourselves and behave responsibly. By 'responsible' is meant that we reduce our emissions, preserve our resources, make our installations and appliances as energy efficient as we can, operate them at optimised levels, and develop them with a view to getting more out of them for less. And this would appear to make sense.

But does it? If we live in a world which is going to grow – in terms of population, in terms of the quantities of what we need, in terms of overall wealth, in terms of energy demand – and grow substantially at that, then is slamming on the brakes and keeping our development as low as possible really the only, and if not the only then the best way forward? Bear in mind how *unequal* our world is today: how much those who have and use energy have and use it, and how little those who don't don't. And how much those who have energy and use it benefit from it, and how much those who don't have or use it suffer from not being able to. Is it really sensible at this juncture, when there are still so many ills in the world which we *know* we can solve and cure and alleviate with wealth, science and technology, that we propose to grind our development down and preserve *energy?* When energy is the first thing you need to develop? And when energy is all around us, in *abundance?*

Could it not be that if we want to get out of this predicament, and start finding genuine, lasting, *era-defining* solutions, we may have to do *the opposite?* It would be a drastic thing to do. It would mean immediately abandoning all talk of 'saving' energy and making deliberate, positive strides towards using *more.* Not wantonly, not for the sake of it, but pointedly, for the sake of economic development and technological advancement.

Would this be suicide? Would it be 'After Us, the Deluge'[LVI] on a grand, global, cataclysmic scale? Not if we tap into this extraordinary wealth of pure energy we have at our disposal. Not if we combine economic development with social responsibility, and technological advancement with environmental stewardship. There is nothing wanton about taking a stance for civilisation. Nor is it random, or opportunist. The fact alone that there is energy in abundance does not mean that we *have to* avail ourselves of it. No, the reason we need to do so is that we want to understand ourselves as members of meaningful civilisations. And there can be no doubt that we do: the violence, the misery, the suffering, the hunger, the disease, the squalor, the hopelessness and the sheer *boredom* that goes with not being part of a civilisation is a wholly unbearable prospect for most conscious human beings; if it weren't so there *would be no* civilisations. And so we have to acknowledge, and work with, three plain observations:

1) **WE WILL LEAD CITY LIVES**   There is a lot to be said for the country, and pleasant it may be for those who can and wish to afford it, but the predominant mode of existence in the imminent and mid-term future is urban. Since 1960, the percentage of people living in cities has gone up in an almost steady curve, breaching 50 % for the first time in 2008 and increasing by just under half a percentage point every year to reach close to 54 % by 2015. That's across the globe. In some parts of the world, both in emerging economies, such as Brazil (86 %), and in mature and leading ones, such as the USA (82 %), the percentage is much higher.[LVII] And even what many of us in Europe or North America, for example, would regard as country living is in fact nothing of the sort: it is city life, set amongst the greenery and gardens of sub- and sub-suburbia; by high-speed rail link connected with, by super-fast broadband and omnipresent telephony tapped into, the city.

And there are good reasons for this. It's in the city that contemporary life culminates and fulfils its potential, it's where commerce, education, culture, service provision and social exchange take place. The fast and accelerating emergence of mega-cities since the mid-20th century is not a function of nature, it's a phenomenon of culture. And so, all their problems, all their vast iniquities and sheer gargantuan challenges notwithstanding, they, like all cities, are an expression of civilisation. We can and need to embrace this, because no matter how much we love the countryside, and no matter how important the country – without a shadow of a doubt – is, for any number of reasons and to an even greater number of purposes, it is urban living that will define this century.

The very definition of what constitutes a city is, of course, debatable. It would probably go too far for us here to get into the discussion of what makes a city a city and how a Western European city, for example, may differ from a South East Asian one, but the point is not whether we live in definable cities or not, the point is that lifestyles which are defined by education, opportunity, economic mobility, communication and cultural exchange are, in essence, not rural but urban, whether they are practised in the city or not. And it is information technology and energy that allows people to lead urban lifestyles, irrespective of where they live geographically.[LVIII]

2) WE WILL GROW   The current United Nations forecast predicts that the world's population will reach 10 billion in the early 2050s.[LIX]

By comparison, in 1804, when Napoleon was proclaimed Emperor of the French, Thomas Jefferson won a second term in office as President of the United States, New Jersey became the last northern state to abolish slavery, and the German pharmacist Friedrich Sertürner first isolated morphine from the opium poppy, the world population for the first time reached 1 billion people. Just over a hundred years later, by 1927, the year *The Jazz Singer* opened in New York as the first feature-length film with synchronised sound (the first full-length 'talkie'), it had doubled to 2 billion. Barely fifty years later, by 1974, when ABBA won the Eurovision Song Contest with *Waterloo,* it had doubled again, to 4 billion. Between Napoleon's and ABBA's Waterloo, the world population doubled twice. As growth is currently fairly steady at around 1.1 % per year, we can expect to reach another doubling – the 8 billion mark – by around 2023.[LX]

These people need to be fed, sheltered, educated. They need to have prospects for themselves and their offspring, they need to be able to see, pursue, and find a purpose in life. They are human beings with the same hopes, the same dreams, the same aspirations and the same rights as us. Are we to deny them their life chances by saying, 'you can't have transport, you can't have health care, you can't have schools and you can't have all the things that make our lives safe, comfortable and convenient and that free us up to balance our time with leisure, pleasure and recreation?' Are we in fact to say to them, 'civilisation is not for you, be you stuck somewhere in a standard of living and an outlook that corresponds roughly to our Middle Ages?' Clearly not. What we are to do is to find a way of tapping into the abundance of energy that we have on our planet and make it available to everyone.

Whether we, from our perspective, which is by necessity culturally biased, think that a farmer living and working the land in a developing country is basically happy and has everything they want and need or not, or whether we think that they should develop and adopt what used to be called 'Western Lifestyles' is immaterial. The fact very simply remains that without energy you can't have state-of-the art medicine, up-to-date education or contemporary means of communication. And these are the things that not only we want, these are the things that manifestly people in every part of the world, almost irrespective of cultural context, want.

And why do people want these things? Because they are fashionable and trendy? Because other people have them? Because there is nothing better imaginable? No, because they make a profound difference to their quality of life and to their outlook.

3) WE ARE IN THIS TOGETHER   We are already connected, and our level of connectedness is only going to increase. The network is the defining structure of our 21st century existence. And this, put simply, means power. In both senses of the word: in the network we have and share power, as in 'the ability to make decisions and exercise control over our lives', and in the network we have and share power, as in 'energy'.

Energy is the lifeblood of what we understand as civilised living. If you are in a position to read this book, be this in print, on a reading device or online, then most likely will apply to you what applies to the vast majority of people in the developed world: a power cut of about ten minutes is a minor inconvenience. A power cut of ten hours will seriously put you out and register as an incident of regional importance. A power cut of ten days is a national emergency. Lives will be lost. So, no: turning back is not an option. Reversing our development, whether this be to pre-digital, pre-industrial or prehistoric days is not worth considering. And nor is restricting ourselves to our current status quo. Nor, for that matter, is looking at energy as an issue of 'autonomy', as something that we can address and handle locally, just for ourselves on the spot where we live. None of these offer a way forward, they all amount to regression, at vast human cost. What is worth pursuing, what will make it possible for us to very soon think in terms of abundance, what will free us up from resources and from the locality where we happen to be, what will lift us onto the next energy plateau, is the network. Because in the network, it doesn't matter that the sun doesn't shine in one place for 24 hours. The sun always shines somewhere. There's some wind blowing somewhere, always. The tides keep coming in and rushing out, all day long, and all through the night. Inside the earth, there is permanent, lasting heat.

There is energy all around us, constantly. Looked at as a whole, earth is a ball immersed in, emanating, spinning on energy. The idea that we have to scratch earth and scuttle around on it, burning up part of it to get at energy is, by now, ludicrous and pathetic. The energy that we need is all, and always, there. So as soon as we feed that energy into a global network, within which it can travel halfway round the globe in a fraction of a second, the local weather, the time of day or night, even energy storage, all cease to be an issue: 'So what if my corner of the world doesn't yield very much energy right at the moment, there's another corner that most certainly does...'

A QUICK REFRESHER ON ELECTRICITY

IV

We've stressed this point already: it matters a great deal how we look at energy, as what. And so it matters as much in what form we handle energy. If we are going to stop looking at energy as 'stuff', as resources, then we can't continue handling it as if it were 'stuff'. That's why electricity is so important. It is impossible to start thinking about, and handling, energy differently, without getting to grips with what it is we're actually talking about, at least at a basic level.

We suggested earlier on that electricity is *abstract* energy. And if you've read the Foreword, you may remember us saying that energy, in the context of information technology, is an 'intellectual construct'. This may sound alienating, but it needn't be. The fact that it's abstract does not make it any less real. The most common forms we physically experience energy as are heat and motion (and light, as it happens, though we tend to take that for granted). Heat we can easily convert into motion, and motion we know how to convert into electromagnetism: electricity. What happens when we do this is of great significance to what we can do with energy next. Because at the point where energy is converted into electricity, it becomes medial, meaning it becomes capable of carrying and of representing information. This is really different to what it does when it's heat or motion. Heat is only heat, nothing else. It cannot by itself convey anything other than what it is. A pot of water can be more or less hot. It can't, in that sense, be 'meaningfully' hot. Similarly, motion is just motion: a wheel can turn faster or less fast, in one direction or another. It can't turn in a way that communicates anything beyond the fact that it's turning.

Yes, we associate meaning with energy even in these basic forms. Fire, the sun, wind: they are all laden with symbolism that we have attached to them. But you cannot imbue the energy that's in a log of wood with information. What you can do is agree with your neighbouring tribe that a fire on the hill conveys a certain message, and you can create smoke signals, true, and we'll be getting on to this and the different levels of symbolisation of energy a little later in this book, but for the moment suffice it to say that it is electricity that you can give explicit, encoded information to, which it carries at extraordinary speed and with near-absolute precision. Put the heat of your log of wood into steam and with that steam turn it into motion that drives a dynamo and you have a form of energy that can communicate all the knowledge in the world. And store it. And send it across the planet, in the blink of an eye. That has never been the case before.

Think of just how amazing this really is: in many countries, Switzerland being a fine example, when you go to a train station, you notice the architecture, perhaps, and most likely the trains. The big, elegant locomotives and the many long modern carriages. Most of them nowadays have air conditioning, subtle lighting and sockets for your laptop. But do you notice the energy? Do you see any evidence of what it takes to shift all that metal, with all those people in it, and in such style, at a hundred, in France routinely at two hundred miles an hour?

No. What the infrastructure provides the train with is a cable no thicker than your little finger. No mess, no smoke, no lorries, no stores, no trailers, no tanks, no noise. Just a cable and a pantograph atop the locomotive to pick up the power. If it weren't human ingenuity, it would be a miracle.

The discovery of electricity – and it was a discovery, more than an invention – was momentous. Up until then, if you wanted to relay motion across any distance – a factory floor, for example – you had to use gears, shafts, axles and straps. Now, with electricity, you were able to not just transport energy without the aid of mechanics across hundreds of miles, you could also easily branch off parts of the energy to another destination and make little portions available for different uses in different parts of the world. *All* different kinds of uses in *all* parts of the world. All you need is a bit of cabling and you can do a myriad of things in any corner of your room, in any corner of the world.

No wonder then we have, after little more than a century, five hundred billion electrical devices.[LXI] And what we also have, for the first time, is something we can call ubiquitous energy: today, energy, abstracted from object, space and geometry, is simply there, wherever we want it to be. So yes: electricity is extremely cool.

But one distinction does need to be made here. We say it was a 'discovery', not an 'invention'. That is certainly the case, electricity was not 'invented' by humans. But nor was it lying around the countryside, tucked away under rocks or perching upon pine trees, waiting for us to come along and pick it up and scrutinise it and exclaim: 'How wonderful! Here's some electricity, let's have it and make good use of it!' We had to go beyond recognising it, we had to learn to understand it, and that meant *thinking* it through in a very specific, particular, mathematical way. Only by thinking it through in the right way could we use it. Bear that thought in mind, and should you ever get a feeling that talking of 'energy through intellect' or 'genius power' is nonsense, retrieve it for a moment. Electricity *in itself* is a great example of how it is the power of our minds that allows us to tap into the power that's around us.

But what exactly *is* electricity? You may know or remember this from school, or you may have studied it and be an expert, and if that's the case and you don't want to read the next few pages, by all means feel free to skip them. But in case you don't, or if you want to refresh your memory: here's a low-down of what we're dealing with.

Electricity is described by five basic values or entities:

A) THE ELECTRIC CHARGE (SYMBOL Q FOR 'QUANTITY')   The generally accepted way of describing electricity runs as follows: atoms consist of a nucleus that is surrounded by electrons. The nucleus of an atom in turn consists of protons and neutrons (there is one exception to this: the hydrogen-1 atom, which has no neutrons).

Protons have a positive electrical charge and neutrons have none. The electrons that surround the nucleus are negatively charged. What holds the atom together is the fact that the positively charged protons and the negatively charged electrons attract each other by what's called the electromagnetic force. You could say that it's this force, electromagnetism, that holds our world together at atomic level.

The positive charge of protons (+1) and the negative charge of electrons (-1) is of the same strength and referred to as the *elementary charge*, which is symbolised by the letter 'e'. The charge is very small. If you multiply the elementary charge by six quintillion and roughly two hundred and forty quadrillion, or 6,240,000,000,000,000,000 ($6.24 * 10^{18}$) you get 1 coulomb (symbol C), which happens to be the amount of electric charge that is transported in one second by a steady electrical current of 1 ampere, about which more in a moment. (To put this in context: a domestic fuse tends to allow for currents of between 5 and 30 amperes.) *Coulomb* thus is the unit for electric charge.

We're all of us familiar with the phenomenon of an electric charge being released, for example from putting on a jumper which makes your hair stand up, because in the action of pulling it over your head it briefly brings the electrical charge of one object (the jumper) in close contact and friction with another (your head), which causes the electric charge in both to be temporarily rearranged.

One way of looking at electric charge might be as the amount of water that you can scoop up with your hand: if your hand is fairly big, the amount of water will be more than if you're a child of five, for example, and your hand is very small.

B) THE ELECTRIC VOLTAGE (SYMBOL V FOR 'VOLTAGE' OR E FOR 'ELECTRO-MOTIVE FORCE')   Voltage describes the amount of energy that is necessary to move an electric charge within an electric field. If you think of electric charge as the amount of water that you can scoop with your hand, then the voltage is the amount of energy that is required to lift the water up within the gravitational field, so it can then be allowed to descend again. The higher you lift the water, the more energy you need to do so. Consequently, from a great height (and therefore with a lot of energy), the water will come down to earth at a great speed and with great force: this would correspond to a high voltage. Whereas from a low height (and with little energy), the water coming down would not have much force by comparison.

So voltage can be described as the capacity of an electrical charge to 'do work'. An electric generator therefore is in essence a charge pump, comparable to a water pump. The unit for voltage is *volt (V)* and the voltage you draw from your power socket will depend on where you are. The most widely used voltages are 100V (Japan), 110V (Belize, Taiwan, Cuba, several Caribbean islands), 120V (USA, Canada, several South American countries), 220V (Argentina, Brazil, Bangladesh, some Eastern European and African countries), 230V (Australia, New Zealand, Europe and many African countries), 240V (Lebanon, Kuwait, Kenya, Liberia and other parts of Africa and the Middle East), but there are also countries that use 115V (Trinidad & Tobago) and 127V (parts of Brazil, Libya, Mexico, among others). The most commonly used triple-A (AAA) and double-A (AA) batteries come at 1.5V. A train, by contrast, will be served by anything between 750V (in the South East of England) and 25,000V (in the rest of England, northern France and large parts of Eastern Europe), with various voltages in between for other parts of Europe.

c) ELECTRIC CURRENT (SYMBOL I FOR 'INTERNATIONAL AMPERE' OR ALSO, LATIN, 'INFLUARE' - TO 'FLOW IN')   The electric current indicates how many electrical elementary charges are being moved through an electric field within a given time period. Applied to our example of water, it describes whether you let the water run through your fingers quickly or slowly. The unit for electric current is *ampere (A)*. Much as an excess of water suddenly pushing through a water pipe onto a piece of equipment might damage it, so a sudden rush of electricity going through a power line into an electric device can be dangerous, which is why all households, and in some countries sockets and plugs too, are fused. As mentioned above, the majority of domestic fuses are specified to between 5A and 30A, meaning that if for any reason the current exceeds that level, the fuse will 'blow' and the supply of electricity to the relevant part of the household or device will be interrupted.

d) ELECTRIC POWER (SYMBOL P FOR 'POWER')   Electric power describes the rate at which electrical energy is transferred in an electric circuit. In our example, this would be equivalent to the amount of water that arrives on the floor at what speed per unit of time. So I can either let a lot of water descend to the ground from a small height, or a little water from a great height: the amount of power I get is the same. The unit for electrical power is watt (W). *Watt (W) = volt (V) * ampere (A)*, an equation that otherwise gets expressed as $P = V * I$ (power equals voltage times current).

A standard old-style Edison light bulb would have used 60W, whereas an energy-saving light bulb giving off the equivalent amount of light will use about 11W. An electric oven will run at around 1,500W, whereas an LED light may use as little as 6W or less. A nuclear power station may produce between one and two billion watt or 1-2 gigawatt.

Electrical power could also be likened to the amount of force a cyclist needs to keep their bike running at a certain speed.

e) ELECTRIC ENERGY (SYMBOL E FOR 'ENERGY')   Finally, electric energy describes the amount of power that is exerted over a certain amount of time ($E = V * I * t$ or energy equals voltage times current times time). The units used are the *watt second* or the more familiar *kilowatt hour (kWh),* which costs about 18 Euro cents if drawn from an average European power socket, or the *joule* which correspondingly also comes in *kilojoule* and *megajoule.*[LXII] One watt second equals one joule, and since there are 3,600 seconds in one hour and a kilowatt is one thousand watt, one kilowatt hour (1 kWh) equals 3.6 million joules or 3.6 megajoules.

A way to illustrate electric energy might involve the aforementioned bike ride. A long bike ride uphill is exhausting because it uses up all your energy, whereas a short trip on a flat stretch of road doesn't require much energy at all. Or we can look at the amount of battery power that we need to illuminate a small light bulb. For example, a 3W-h battery is capable of keeping a 6V/0.5A bulb lit for one hour: 6 (volts) × 0.5 (amperes) = 3 (watt-hours). So, if, instead of using a 0.5 A lamp, I use a ten times more efficient 0.05 A LED-lamp, the same battery will last me ten times as long: 6 × 0.05 = 30.

Here, for comparison, are some energy values:
0.000 000 001 J (nanojoule, nJ)
One billionth of a joule. 160 nJ is about the amount of energy a mosquito uses to stay in the air.
0.000 001 J (microjoule, µJ)
One millionth of a joule. At the Large Hadron Collider (LHC) near Geneva, particles collide with approximately 1 microjoule of energy.
0.001 J (millijoule, mJ)
One thousandth of a joule. Typing a letter on your computer keyboard requires very roughly a millijoule of your energy.
1 J (Joule, J)
One joule. About the amount of energy it takes to lift an apple up by one metre.
1,000 J (kilojoule, kJ)
One thousand joules. About the amount of energy that one square metre of the earth receives in solar radiation in one second.
1,000,000 J (megajoule, MJ)
One million joules. About the amount of energy it takes a one ton vehicle, like a car, to move at 100 miles per hour. 3.6 MJ is the equivalent of one kilowatt hour, which is the standard metering unit for electricity.

An adult woman weighing roughly 70 kg uses about 6.3 MJ on an average day, without carrying out any special exercise, and one kilogram of coal, when burnt, releases just under 30 MJ.
1,000,000,000 J (gigajoule, GJ)
One billion joules. A barrel of oil, when burnt, releases approximately six gigajoule.

86.4 GJ is the same as 24 megawatt hours (MWh). This is roughly the amount of energy that a standard-sized 1-gigawatt power station generates in a day.

1,000,000,000,000 J (terajoule, TJ)

One trillion joules. One square kilometre of photovoltaic solar foil surface can collect about 360 TJ in a year, if exposed to average levels of sunshine in Central Europe.

1,000,000,000,000,000 J (petajoule, PJ)

One quadrillion joules. The standard sized one-gigawatt power station mentioned above puts out approximately 31 PJ in a year.

In 2009, the whole of Switzerland used approximately 877 PJ of energy.

1,000,000,000,000,000,000 J (exajoule, EJ)

One quintillion joules. The United States in 2007 used about 94 EJ. Switzerland receives about 200 EJ of energy from the sun in a year, while the entire global energy use in 2007 amounted to about 503 EJ.

1,000,000,000,000,000,000,000 J (zetajoule, ZJ)

One sextillion joules. Earth receives about 10.7 ZJ of energy from the sun each day. Global annual energy consumption lies at around 0.5 ZJ.

1,000,000,000,000,000,000,000,000 J (yottajoule, YJ)

One septillion joules. About the amount of energy it takes to heat all the water on earth by 1 degree Celsius. Earth receives about 3.9 YJ of energy from the sun each year.

You'll remember us talking about 257 Pearl Street, where Edison built his first commercial power station generating electricity for lower Manhattan. In it, Edison operated 6 'Jumbo-Dynamos', which he developed himself and which at the time were the biggest of their kind. Each weighed 27 tons and produced 100 kilowatts, or the amount of power required to feed 1,200 electric lights at the time. While dynamos – electricity generators – have become bigger since then, and somewhat more efficient, the principle at work in them has hardly changed at all, and today's generators still turn energy into electricity much in the same way.

I   THE DYNAMO/GENERATOR

What actually happens in the dynamo or generator? Fire produces heat; heat gets water boiling; water that boils evaporates into steam. The steam, when funnelled onto an appropriately shaped wheel, sets in motion a turbine, and the mechanical energy of this motion is then converted into electrical energy. This much pretty much everybody knows. What happens inside the generator, though, still has something of the mystical about it, even though it's so commonplace that you can fit it to a bicycle wheel:

If you turn a spool of copper wire inside an iron ring that has also been wrapped in copper wire, then the force of electromagnetism causes electric energy to be generated. The phenomenon, or effect, is called electromagnetic induction and, unless we were to go into great detail on electromagnetism, that is the simple and almost infuriatingly straightforward answer. It's just the way it is. Metals have magnetic fields, and if you rearrange their magnetic fields (by moving them in close proximity to each other) you also, and always, rearrange their electric field. The energy you put into rearranging the magnetic field (the movement) goes into the rearrangement of the electric field (electricity).

The astonishing and immeasurably handy thing is that this energy can then be channelled through a wire across almost any distance, and at the other end of the wire a near-identical set-up – a rotor or electrical motor – can reconvert the electric energy back into motion, and any variety of alternative set-ups can convert the energy into light or heat or sound, or practically anything else you care to think of. What makes all of this even more remarkable is that instead of having electricity flow from your dynamo to your rotor at the 'back-end' of your system, you can, while leaving everything in place, also turn the rotor and you'll find that the dynamo at the 'front-end' now starts to turn: the energy travels in the reverse direction just as easily. Yet for all this astonishing versatility, all it takes to generate electricity is two types of metal, arranged in one particular way, and then set in motion. That's all. You can try this at home, it will work…

II   AC/DC

Another thing that is relevant to our subject of electricity, while we're on it, is the difference between alternating current (AC) and direct current (DC). The reason this is relevant is twofold: firstly because AC and DC have very different qualities when sent over long distances (DC being far more efficient), which is something we can expect to be doing more and more as electric power networks proliferate and grow larger, and secondly because photovoltaic solar power as well as any battery, both of which we expect to be using more of, produces one (DC), when the entire user supply of the developed world is geared towards the other (AC).

Edison's first power station generated and distributed electricity as direct current, but by the end of the decade (the 1880s) Edison's arch rival, the entrepreneur George Westinghouse,[LXIII] prevailed with a competitive system that used alternating current instead. The advantage of alternating current over direct current is that it can easily be up-converted to higher voltages, which present much lower energy loss over a distance than low voltage systems. So while, for example, a typical high voltage power line at 110 kilovolts (110,000 volts) shows a loss of power of 6% over 100 kilometres, at 800 kV (800,000 volts), this drops to 0.5% over 100 kilometres. If, by contrast, you were to send electricity down a cable at a domestic 230 or 120 volts, then after 100 kilometres there would be hardly any power left at all. The highest voltage cables now being developed use 1,300 kV or 1.3 million volts, with energy losses as low as 0.3% over 100 kilometres, and at these sort of voltages the advantage of DC's greater efficiency outweighs the disadvantage of its more difficult conversion from higher to lower voltages, which is why very high voltage power lines tend to use direct current rather than alternating current.

Today, most domestic and commercial electricity grids use alternating current, while some industrial and urban public transport systems (the London Underground, for example) use direct current.

With voltage being such a significant factor in determining the distance over which electricity can usefully be sent, we should take this opportunity to look at one more piece of the jigsaw:

III   THE TRANSFORMER

A transformer, much as the name suggests, transforms an electric current, specifically an alternating current (AC), from either a higher voltage to a lower one, or, inversely, from lower voltage into higher. This is useful in any number of electrical devices and installations; it is what makes electricity safe and economical to use at domestic level, because voltage is sufficiently low, while it also allows power grids to send electricity over long distances, because as we've just seen only at high voltages does it travel with acceptable efficiency.

There is a vast array of different types of transformer, ranging from tiny ones in small portable appliances to enormous ones at giant power stations, but in the majority they all work along the same principle. What you need to make a transformer is:

A magnetic core

This is normally a piece of metal, such as 'electrical steel' (also known as lamination steel, silicon steel or transformer steel). It's a steel that has particularly suitable electromagnetic qualities and therefore lends itself especially well for the purposes of a transformer. Cores made of any type of steel are known as ferromagnetic cores (from Latin *'ferrum'* – iron). They are by far the most widely used. But other types of core, such as ones made of ceramic, or ones that actually haven't got a core and are therefore known as 'air cores' are also possible. The core normally has the shape of a square O, though it can also be circular.

A primary coil or winding [NP] and a secondary coil or winding [NS]

This is any type of wire (often copper, for its well known conducive properties) that winds or coils around the core, in such a way that the two coils face each other, with the hole of the O in-between them. The two coils need to be separate from each other, so you're really looking at two wires.

An electrical current [VP] that goes into the primary coil [NP] and an electrical load that causes an electrical current [VS] to come out of the secondary coil [NS]

You need an electrical current that goes into the primary coil NP and you need something connected to the secondary coil NS that requires an electrical current (like a machine or a light or a grid, anything that uses electricity), in order for it to flow.

What happens when this set-up is in place is that the electrical current that goes into the primary coil (NP) creates a magnetic field inside the core. This magnetic field induces an electrical current in the secondary coil (NS), which is what is meant by 'magnetic induction'.

Now, if both the primary coil (NP) and the secondary coil (NS) had the same number of turns, then the voltage induced in the secondary coil would be exactly the same as the voltage in the primary coil. (It would be, that is, if the transformer were 100% efficient, which is not possible; but a good transformer has about 98% efficiency, so for practical purposes the loss of voltage is, if not negligible, so certainly very low.)

This would not, however, make for a very useful transformer, since the output would be the same as the input, and the installation would therefore be pointless. But all you need to do to actually transform the electrical current is make sure that the number of turns in the primary coil is not the same as the number of turns in the secondary coil, and what happens then is that the voltage changes in direct proportion to the number of turns in each coil: if you have more turns in the secondary coil than in the primary coil, the voltage you get from the secondary coil will be higher than the voltage you have in the primary coil. If, by contrast, the number of turns in the secondary coil is lower than the number of turns in the primary coil, then the voltage you get in the secondary coil will be lower than the one you have in the primary coil.

So say you wanted to transform a voltage of 120V to 12V, then you would know that 120 is 10 times 12, so you could give your primary coil 1,000 turns and your secondary coil 100 turns. 1,000 is 10 times 100, so just as you'd expect, your voltage in the secondary coil would be 10 times lower than 120V = 12V. Result.

What you're actually doing is transform a current that has 0.1 amperes, 120 volts and, for the sake of argument, 12 watts power into one that has 1 ampere, 12 volts and still produces 12 watts of power. Note that the current (the ampere unit) goes up by a factor of 10, whereas the voltage (the volt unit) goes down by a factor of 10, while the amount of power (the watt unit) stays the same. (In Europe, you'd be more likely to want to convert from 230V to 12V, which is just as simple, if not quite as elegant, since 230 happens to be 19.2 times 12, so if you have 1,000 turns in your primary coil you'd need 52 turns in your secondary coil.)

Since the process is entirely reversible, you do exactly the same thing in the reverse order to get the opposite result. So if the solar foils on your roof produce electricity at 12V (as they are quite likely to do), then it will be your primary coil that has 100 turns and your secondary coil that has 1,000 to get the voltage to the level where it can be fed into the grid. (That's in the US, for example. In Europe, it would once again be 52 and 1,000 respectively.) What you'll also need to do is convert the direct current into alternating current, for the purpose of which you'd use an inverter, which does both the transforming of voltage and the conversion of current at the same time.

It was the English chemist and physicist Michael Faraday [LXIV] (he of Faraday Cage fame) who first came up with the idea of electromagnetic induction in 1831, or rather: it was he who first published his findings. The American Joseph Henry [LXV] was working on the same thing in the same year, but only got to publish his ideas later, so he rather lost out on the credit for this splendid invention that we pay little attention to but that makes much of what we do today possible.

In fact, Faraday and Edison, Alessandro Volta [LXVI] (who is credited with inventing the battery in 1800, and after whom voltage is named), André-Marie Ampère [LXVII] (who was instrumental in discovering electromagnetism in the early 1800s, and after whom the unit of electric current is named) have, with their work, changed practically every single aspect of our lives and propelled the Second Industrial Revolution into the 20th century as technology became the central determining factor in our development, with instantaneous long-distance communication, public transport and mass entertainment starting to feature and co-determine how we experience our existence on this planet.

From Edison's lone, coal fired power station in Pearl Street, we have moved on to a world of tens of thousands of electricity generators of all different types and sizes. As of 2014, coal-fuelled power stations accounted for about 39% of all the electricity generated in the world, gas (22%) and oil (5%) together made up just over a quarter (27%), whereas wind (3%) and hydropower (17%) combined came to about a fifth or 20% of all the electricity we produced, and nuclear power to about 11%. The remaining 3% approximately was generated by tidal, wave, geothermal, biomass and solar. [LXVIII]

The largest hydroelectric dam in the world today is still the Three Gorges Dam in China, which is capable of producing up to 22.5 gigawatt (a standard-sized large dam such as the famous Hoover Dam produces less than a tenth of that, about 2 gigawatt, which in turn is roughly the same as an average nuclear power station). [LXVIX] And while wind installations make an as yet relatively small contribution to the overall pie, their number is growing fast and by 2016 they were able to contribute nearly 5% of global electricity demand. [LXX]

What all these many varied installations have in common is that they almost all use the almost exact same simple science that took us no more than a few pages to explain to turn motion into electricity, thanks to the phenomenon of electromagnetism. What distinguishes them from each other is their size, their location and the source of the motion that drives their generators. Some use nuclear fusion to generate heat, which creates steam that drives the turbines; some use coal, oil, gas or biofuels to do the same thing; some take the motion from water pouring down shafts in dams or flowing over turbines in rivers; and some take the motion from rotor blades that turn in the wind. The only exception to this is photovoltaic solar, about which we'll be talking quite a bit more in just a moment.

All of it serves the very same end: to keep us supplied with electricity. And electricity we certainly want. In 2013, world average electricity consumption stood at 3,104 kilowatt-hours (kWh) per person, and rising. But there are vast variations from one country to another. The average man, woman and child in Ethiopia, for example, in 2013 used 65 kWh, which is the amount of energy you need to keep a single 60 watt light bulb burning for three hours every day. The Nepalese used about twice as much, whereas in India the figure was 765 kWh, and that's bearing in mind that in the same year in India alone 250 million people still had no access to electricity at all. The average Brit in 2013 used 5,407 kWh (a decline from 6,061 kWh in 2008), the average German a little more at 7,019 kWh (remaining steady), and the average American almost twice as much again, 12,988 kWh. Kuwait reached close to 15,000 and Norway just over 23,000 kWh.[LXXI] Iceland easily continued to top the list with 54,799 kWh, though this figure is distorted, since Iceland has a lot of extremely cheap geothermal energy and hydropower that it uses to manufacture aluminium for export, which is one of the most energy-intensive industrial processes you could choose to get involved with. Iceland is so rich in easily accessible geothermal energy that it can afford to heat pavements in winter so they don't freeze over and make beaches and sea water pleasant and warm, because the energy is all there to be used. It is, as it happens, an example of a country with abundant energy...

V   WHERE MUCH OF IT COULD BE COMING FROM INSTEAD

Now, if we wanted to produce *all* of that electricity, generated by hundreds of nuclear power stations, tens of thousands of coal-fired power stations, including even the many thousand water, wind and geothermal power plant, from photovoltaic solar cells, it would take approximately 500,000 square kilometres of surface area to do so. This does sound like an awful lot. But have a look at a map of the world and pick out Spain. 500,000 square kilometres is about the size of Spain. You can also pick out Texas. It's about two thirds the size of that. Or the Gobi Desert in China: about half of that.

So if it were possible to produce flexible photovoltaic solar cells that can be manufactured in great quantities very cheaply, that can be formed into any shape, used in any type of location, that can simply be put there and wired up to provide energy, very much on a 'plug-and-play' principle, on a scale from a few watts, like a small battery, to a few gigawatts, like a nuclear power station, then we would view the role that solar energy can play in our electricity equation with fresh eyes. But the all-important thing would be to connect these solar cells and let them feed into a massive, global network, so that, overall, the energy supply would be constant and guaranteed, so that individual users or regions or countries would not suffer from the fluctuations in weather, season or time of day and night: so that the energy supply would be *always on.* A bit like the internet where we have constant access to information, except what would be exchanged, shared and traded here would not be information, it would be energy.

This is now possible. And there's a compelling thought we want to end this section on. It illustrates in the simplest way the principle of 'always on' energy. We can now make, for a few cents or pennies a piece, flexible, thin foils of semi-conductor material. We can simply determine the properties of the material: it can either collect light and convert it into electricity, or it can as easily do the opposite, convert electricity back to light. It can, therefore, be either a photovoltaic solar cell or an LED lamp. The weight, cost, dimensions and the technology at work are all the same. So we can make two foils that look, feel and weigh the same, and connect them with a thin wire. One of them we can put outside, facing the sun, the other we can keep indoors. As long as one is facing the sun, the other one will light up. That's really as simple and straightforward as it is. We can set these two foils hundreds, indeed thousands of kilometres apart, and the same will still be the case: as long as one is facing the sun, the other one will be lit. And this means that as long as the sun is shining somewhere on earth, there can be 'light' somewhere else. For 'light' read 'power'.

We will be talking some more about photovoltaics; what it takes to make photovoltaic solar cells, how they work and how, unlike any of the other electricity generating methods described or mentioned above, they convert solar energy directly into electricity, without going through any mechanically induced electromagnetic process. Also about how, once they are made, they can garner energy without causing any pollution, any byproduct or any waste; and how manufacturing them, although at the moment still complex and expensive, is becoming cheaper every year. How photovoltaics, for all these reasons, open up a whole new ball game that is manifestly different to any that we've been playing so far. But before we do so, let's examine what's so special about the *network...*

THE NETWORK: SHARING POWER

V

On 23rd April 2005 (coincidentally William Shakespeare's 441st birthday), at 8:27pm local time, a young man called Jawed Karim [LXXII] uploaded a video clip of 19 seconds duration onto a server in California, entitled *'Me at the Zoo'*.[LXXIII] In it, Jawed stands in front of the elephant enclosure of San Diego Zoo, speaking to a camera (held by a Yakov Lapitsky,[LXXIV] a high school friend of Jawed's), uttering the insightful words: "Hi. So here we are, in front of the elephants. The cool thing about these guys is that they have really, really, really long, erm, trunks, and that's... that's cool." Following a short pause during which he looks back at the animals so described, he turns around to camera once more and concludes: "And that's pretty much all there is to say."

Now, elephantologists the world over may cavil with Jawed's somewhat hasty conclusion, and admirers of The Bard may consider his way with words a touch prosaic, but one thing Jawed and his mates Chad Hurley [LXXV] and Steve Chen [LXXVI] had got spot on was their assessment of their fellow human beings' eagerness to share videos they'd made themselves. YouTube was born.

Within a few years, YouTube had established itself not just as a handy video-sharing website, but as an iconic brand, a social phenomenon, a highly-prized asset to its subsequent owner Google and as a byword for user-generated content and a new video culture. And this could well be what matters most. With YouTube, two things to do with video radically altered, almost overnight: 1) the quantity of content that's available online for free, and 2) the relationship between us and that content.

And what does this have to do with energy? Well: on YouTube you can say without hesitation that there is enough video content. There is more than enough, there is, you could argue, too much. But too much for what? Is there an upper limit to how much video content is 'good for you'? Or is there a practical limit to how much of it we can manage? After all, we have to somehow find our way through all this material. Or do we? Do we have to be able to sort it, choose it, select it, or are we happy to just 'graze' on it, nibble at it a bit when we feel like it and then turn our attention to something else. What about copyright? Who actually owns all that content? That's simple enough when it's just you with your camera shooting your two best friends careering into a wall on their home-made scooter and laughing their heads off. But what if you record strangers? Do you ask them if they mind you putting their image online? What if you put a soundtrack on it. Do you really check back with Tracy Chapman's manager, every time you use *Fast Car?* Everything anybody used to know about how video content was handled, managed, financed, sold, distributed or controlled has more or less gone out of the window, thanks to YouTube. The whole set of rules is still changing, even as we speak, more than ten years later. Because it has to. And because it can. And the rules are largely being made up by 'the community'. Legislators, corporations, politicians, sociologists, even YouTube itself, are all just catching up with the virtual reality that's being created.

So does that make for a better video world we live in? Some would say it does, others would say it doesn't. It absolutely depends on how you position yourself towards it and what you deem 'better' or 'worse'. What is certain though is that it's a different video world we live in. One in which we have a far wider array of choices, not only as consumers, but also as providers. As arbitrators. As communicators. As participants. And remember, they are choices: you don't *have to* post your videos on YouTube, you don't have to comment on them. You don't even have to watch any. There is no law that says you need to engage. The crucial point is that if you want to, you can.

The way we 'play' with video habitually now is unlike anything we'd done before. It's not just that we have more videos than we used to, the whole setting, the whole landscape, has changed.

YouTube is just one impressive example of what happens when users become producers, when content gets shared and when the sharers are networked. Another is Wikipedia. Another is Twitter. Another Facebook. What they have in common is that they connect people, and that as soon as critical mass number of people are connected, user-provider relationships change completely, and this in turn means that radically new user-provider behaviours emerge, with breathtaking speed.

What we are beginning to get our heads around, and what we have the extraordinary luxury of being able to call our 'problem', is that we have to rewrite all the rules. And what at the same time disconcerts us, understandably, is the pace at which things can happen, mostly thanks to the network. And what's even more unsettling, perhaps, is that nobody exactly knows what 'the network' is, not least because there isn't really just one network, there are multitudes of different networks that are enmeshed with each other and that exercise great influence on each other too.

I   **POWER SHIFTS OF A DIFFERENT KIND**

As we are writing this, two seismic shifts are taking place in two very different parts of the world, both geographically and conceptually. The first is the Arab Spring which has been underway since the tail end of 2010, but mostly – and hence of course its name – came to manifest itself in the early months of 2011. The other is the dramatic extraction of UK politics from the grip of Rupert Murdoch's media empire.

We are interested in them here not because of any connection they may have – directly or indirectly, certainly in the case of the Arab Spring – to energy, but because they exemplify how network technology can facilitate rapid developments that do away with an old order and usher in a new order, which is defined by 'the people'.

The democratising effect in both instances is palpable and pronounced. While back in 1992 Murdoch's *The Sun* newspaper was able to trumpet *'IT WAS THE SUN WOT WON IT'*[LXXVII] about the UK general election of that year, which, against all predictions and polls, had narrowly gone to John Major's Conservative Party, by 2011 it was effectively Twitter that brought down *The Sun's* Sunday sister paper, *The News of The World.* Except of course it wasn't *The Sun* that won it for Major, nor was it Twitter that did for Murdoch. It was, in both cases, the people. But back in 1992, Rupert Murdoch's News International, which owned *The Sun, The News of the World* and *The Times* and *Sunday Times,* was fiercely controlled by the mogul, his family and a select coterie of henchmen and women, many of whom have since faced criminal charges for illegal phone hacking and corruption.[LXXVIII] The 'culture' within their news organisation was to bribe, intimidate and manipulate people while inventing, twisting, distorting and concealing, as necessary, facts to the end of, on the one hand, selling newspapers, and, on the other, gaining and exercising influence and power over elected politicians. In 2011, on Twitter, minute-by-minute updates, debates, information, video clips, news stories from around the world, commentary, analysis, and links to petitions and action group websites were freely and openly shared and virally disseminated with nobody having control nor even any steering influence in the matter.

The impact, in both cases, was profound. While it is often maintained that Murdoch's newspapers had far less of an influence on voter behaviour than they themselves claimed and enjoyed winding politicians up with, what is absolutely certain is that politicians of all parties were in awe of Murdoch and his papers, that they deferred to him, or if they didn't, that his papers would set out, and in most cases be able, to damage their reputation and therefore their chances of being elected. Power was therefore held and concentrated very much within Murdoch's circle.

By contrast, Twitter is not always taken too seriously and there are still commentators who think it is not much more than an irrelevance. Yet without Twitter, it would have been unthinkable that two online campaigns demanding for Murdoch's takeover of Britain's biggest broadcaster BSkyB to be stopped would have succeeded in attracting a quarter of a million submissions within 72 hours. This could not be ignored by either the Culture Secretary, who made several references to the unprecedented number of emails his office had to deal with, nor by advertisers, who, one by one, found it to their advantage to announce that in view of public disgust over the Murdoch paper's behaviour, they would for the time being withdraw their advertising from it. Murdoch closed the title, his son James was deposed as Chief Executive of News Corporation's UK newspaper division and the corporation's bid to take over Britain's biggest broadcaster was at least temporarily foiled.

In the case of the Arab Spring, the relationship between social networks, the media and the – over extended periods daily – demonstrations on the streets are fairly complex and may merit a more in-depth analysis than a few paragraphs in a book about energy can provide. But there is little doubt that here, too, it was people's ability to share information about events in real time, to capture evidence about what was happening in their area and publish it for the whole world to see, within minutes, to bypass the dictatorial and monopolist structures set up to prevent them from expressing themselves, that made these rapid and radical changes possible.

The network changes the balance of power between the people who participate in it, it changes the flow and direction of communication, and it changes the role we all play once we are connected. And this *is* very new to us. It's something we are all learning to handle and get to grips with as we go along, and so naturally there are hiccups along the way. Still, in the context of information and data, of media content and knowledge, of news and communication, we are doing our learning very quickly indeed.

Wikipedia, which was launched in January 2001 has, over fifteen years, become the seventh most accessed website in the world, with 40 million articles in more than 250 languages and some 500 million readers globally.[LXXIX] Although it is written and edited by anyone who feels like it, irrespective of their expertise or qualification, it has accuracy levels comparable to those of the long-established and professionally curated *Encyclopaedia Britannica.* But it is free, universally accessible, and extremely fast. When in the evening hours of Tuesday 24th January 2012, news started to spread on Twitter that the highly respected but relatively obscure Greek film director Theo Angelopoulos had died in a road accident in Athens, it took the English language Wikipedia entry less than half an hour to accurately update his biography and reference the date, location and cause of his death.[LXXX]

The power of the network lies in its speed and in its broad base of sources and outlets. It is not, therefore, without danger. In January 2012, Luke Lewis, then Editor of the *New Musical Express,* found himself compelled to issue, via his Facebook page, what he himself called a 'shamefaced' apology. He had inadvertently and with no real ill intentions started what turned into an impromptu hate campaign against a hapless young singer from North Yorkshire whom he didn't rate very highly.[LXXXI] It was an illustration, albeit fairly harmless and apparently taken in good spirit by the young singer, of how quickly things can get out of hand when the power of the network comes into play. By now, both the phenomenon and the expression of something – a video, a picture, a tweet or a blog entry – 'going viral' had entered our consciousness and our vocabulary.

II   THE NETWORK EFFECT AND ENERGY

Networks have become part of our reality, with all their potency and previously unknown dynamics. They are not in themselves good or bad, but they are notably, substantially different to what we've been used to in the past, and they make it possible and necessary for us to learn to handle them. If we handle them well, as the example of Wikipedia, the instances of dismantled dictatorships and loosened grip of dictatorial media barons show, we can genuinely improve the way things are. And, crucially, it is through networks that we can bring together, pool, index, sort and distribute hitherto unheard of quantities of content, without anyone exercising central control. And that is probably why networks matter most. Because imagine this same principle applied not to videos, not to information, but to *energy*.

Today we can indeed expect something very similar to happen with energy. And if our experience with information technology and our understanding, young as it is, of network technology are anything to go by, then we can expect similar effects in principle with energy as we've seen with information:

- A proliferation of sources: instead of having just a few originators of energy (power stations), we will have a multitude of them, of every description and size.
- A dispersal of direction: instead of energy flowing in one direction, from the top (the power station) down (to the consumer), it will flow in all directions, as the roles of 'producers' and 'consumers' become interchangeable and therefore more and more irrelevant. Instead of 'consumers', we become participants and traders.
- And as a result of the above, together with affordable, easy to use

plug-and-play solar capture technology, an *abundance* of energy. None of this is without its own problems, we are aware of that. But it is still extraordinarily exciting. Because this has not been the case before. If we can bring this about – and there is something of an 'if' attached – then we will find ourselves for the first time in our history in a situation where we have not just enough energy but *more than enough.* Many who read this will have their doubts and may even be downright hostile to the idea, and we will be addressing at least some of the questions that we can anticipate a bit further down. But just for the moment we want to allow ourselves that thought on its own, because the development is underway, it is already happening.

We are creating an energy culture that has more similarities with the way we handle, and structure the handling of, information today, than with how we handled and structured the handling of energy over a century ago. And the exciting thing about this is not even just that there is all that energy, the exciting thing is that we are all of us going to have a say in it. And a say in it we will want, because there are big issues at stake. Videos may affect some of our lives some of the time, energy affects every part of our lives, all the time.

Wherever you happen to be right at the moment, you can take a look around you and pick any object, any element of your environment, and ask yourself, what would this part of my existence be without energy. From the clothes you wear to the book or tablet or laptop in your hands; your eyeglasses or contact lenses, if you're wearing them, to the table in front or next to you or nearby; the glass on it, the drink in it; the house where you're staying with its bricks, mortar, wood, glass, furniture, wiring, lighting, kitchen, bathrooms and garden; the car or bicycle outside and the road leading up to the house; the transport and communication connections. Your ability to read in the first place, your ability to go from A to B. Your 6-monthly dental check-up and your hospital treatment if you break a leg or fall ill with cancer, your children's birth and your parents' care in old age, the countryside where you go for walks and the city where you go to concerts, plays, films, football games or meet with friends; the food you eat and the way your waste is disposed of. Everything you've ever bought. All the packaging of everything you've ever bought. Every book, every DVD, every gadget every *thing*. *Obviously* how you keep warm. And how you keep cool. Think away energy from any aspect of anything you are familiar with and you're in trouble. Immediately, within seconds. When we (or anybody else for that matter) come out with a statement like 'energy *matters*', it does sound like a platitude. It is no such thing. Without energy, we are nothing and nobody, leading an existence that to our minds, the way we understand ourselves as human beings, is not worth living, and that, for what it would be worth, wouldn't last very long. If you're over the age of 21 today as you're reading this, you have already outlived your life expectancy in a world without energy: without energy you would almost certainly be dead by now.

Today, we are standing right on a turning point; we can almost sense it as we pivot. We're beginning to not just understand the astonishing power of networks and their dynamics, we are getting used to non-linear, many-to-many structures, in which there is no central control, in which content (information, data, energy) flows from peer to peer, in which the nodes on the network don't just exist and do one thing (for example consume energy), but *behave* (for example consume energy some times, store it at other times, at times feed it back into the network; adjust the amount of energy they use in accordance with what's available in the network and with what the going rate for it is), where they become contributors and therefore direct and indirect manipulators of the system as a whole. In the network, and thanks to the network, we no longer have to accept that anything we do uses *up* energy, that is then no longer there. Instead we can be active in the network and through our own activity provide the potential for further activity. That's what makes it so different, and that's what makes it so interesting, because today, for the first time ever, we can organise not just information, but *energy* in the network.

III   LIFE AFTER FOSSILS

We don't subscribe to the view that just because a lot of people agree on something, it necessarily has to be true. The fact alone that there is no or little dissent does not serve as proof for veracity. But we also don't think it reasonable to disregard overwhelming evidence and broad agreement across wide and varied sections of the scientific community that the answer to the question 'how do we solve our energy problems into the long term future?' is not 'by using more fossil fuels'.

It may neither be possible, nor is it necessary, to stop using all fossils overnight, but when we set about defining a strategy for covering the energy needs of bigger, more sophisticated societies with greater and more persistent demand for power, then it is obvious that fossil fuels will play a diminishing role. The reasons for this have been cited and analysed by many a well-informed person and institution before us and there is no need for us to go into them in detail again here: there are mountains of statistics that prove or disprove, highlight or obscure one fact or another. We are not interested in them. Why? Are we so tied into our world view, into an ideology that we don't want to hear the evidence? Absolutely not. The opposite is the case.

We want to get away from ideology and we do not have a world view to propagate. What we want to do is raise our heads above the parapet and look at things beyond the boundary of human achievement thus far. We want to take a new stance, conceptually. And that means giving fuels, particularly fossil fuels, a very different and much, *much* less significant position in the overall picture.

In the longer term, we can, we may have to, but more to the point we will *want to* get away from fossil fuels pretty much altogether. We will be able to cover so much of our energy demand from other sources, we will not be vastly interested in fossils any more. And for those parts of our energy demand where we don't have any readily available alternatives to fuels, we will be able to produce them in ways that have far less adverse impact on the environment than fossils.

Of most interest to us are solar fuels, in the main hydrogen and hydrocarbons. These are fuels made by a process of either hydrogen electrolysis or methane synthesis. Although their development has some way to go yet, the science behind them has been known and studied for well over a hundred years. Their great advantage lies in the fact that they are carbon-neutral and perennially replenishable, because the elements they use are also abundant (oxygen, water, carbon), and the energy required to put them into usable form (hydrogen and hydrocarbons) is solar.

The potential quantities are remarkable. In Central Europe, it takes about 0.2 square metre of photovoltaic solar cells to get enough energy to produce 1 litre of solar hydrocarbon per year. This means that a relatively small solar farm of, say, 250,000 square metres (about 547 yards in each direction) is capable of producing 1.25 million litres of perfectly carbon neutral fuel. In hotter, sunnier climates, such as that of Saudi Arabia, for example, it would take approximately 400 kilometres squared (or about 7 % of their land surface) covered in photovoltaics to garner enough electricity to produce as much in solar fuels as the country is currently pumping oil out of the ground. So solar fuels hold considerable promise for bridging supply gaps, while allowing us to continue using our existing infrastructure as we gradually adapt it for other types of energy.

That's why we're not interested in whether coal will last us another 100 years or another 200. Whether the Alaskan oil fields might give us oil for an extra 30 years or not. These questions are not irrelevant, but neither are they the big, fundamental questions that matter in the long term. They are interesting in the context only of a transitory period – if things happen fast, about 25 years or so, if they happen more slowly, perhaps 50 – which we'll need from now until we have weaned ourselves off fossils more or less completely.

Aviation and large-scale seafaring are the only two areas that we may not be able, within that kind of time frame, to serve better, more economically, more intelligently and far more plentifully from inexhaustible sources of energy than from fossils. They represent less than 10 % of our total energy consumption, and even for these, we do have, with solar fuels, for example, increasingly viable alternatives.

So by and large, we can safely say: don't bank on fossils. Fossils run out, fossils pollute the atmosphere, fossils are the subject of wars. And this is not polemic, there is plenty of evidence that some of the most catastrophic armed conflicts in our history have their roots entirely in a scramble for control over resources. And it's worth emphasising once more: it's not that we have anything against fossils in principle. But we don't see how they can become unproblematic any time soon. And we don't see them hold any great potential for the future any more, as they did, at the beginning of the 20th century. So fossils, too, are neither 'good' nor 'bad', they simply have played their part and may now retire.

But what *does* come after fossils? Most likely a combination of things. Solar fuels we see as being able to play an important role both in the short and longer term, because they slot in easily and, as the technology develops, increasingly cheaply into the existing infrastructure. But the real step change will not be brought about by converting solar energy into fuels, but by converting solar and other infinitely available energy into electricity and networking it.

IV   'INTELLIGENT ENERGY'

The key to networking energy, as we have seen, is electricity. It's the only way in which it is *possible* to network energy in a meaningful way. We've already looked at how electricity has transformed our lives in a bit over a century. And it's worth emphasising that it has done so not only in quantitative terms, but in *qualitative* ones. Of course, electricity has illuminated our cities, made us travel faster than ever before and enabled us to talk across distance. But these are mere accelerations and enlargements of things we were already able to do: we already had lamps for our cities, only that they were gas lit, we could already travel across the land, only slower and under steam, we were already able to talk, only we had to stand in front of a person in order for them to hear what we were saying. In this context, all that electricity has done is given us more and faster energy.

What is far more significant though is the fact that electricity lay the ground for *intelligent* energy. We should stress here, perhaps, that no computer, no matter how powerful, is as yet truly 'intelligent', but most of us today understand that information technology is different in character from anything we've had before, and here we are getting close to the heart of the matter. Because in what way exactly is 'intelligent' or 'virtual' energy different from what we've had before?

On the surface, and most striking, is the multiplication of functions and therefore uses. You hear us refer a lot to smartphones and the extraordinary scope for versatility that they offer, purely because they aren't portable analogue telephones but small pocket computers capable of running applications. They are such a good example because they show more impressively than any other device we're currently familiar with the difference between *having to do* something and *being able to do* something with a device or machine. An analogue portable telephone, of the kind that we still remember people carrying around in briefcases and later strapped to their belts in the late 1980s and 1990s, you *have to* use in a certain way. There are only so many things the device is ever going to be capable of doing, and if you want to avail yourself of its capability then these are the things you are going to do with it. The moment the device becomes 'intelligent' though, and starts to run applications, it becomes capable of being whatever we want it to be. Far beyond any use we determined for it at the outset, when we invented it, it can now become the device for which programmers and developers around the globe invent perfectly new uses. This is the big shift that has very profound implications, because it changes our relationship with the device itself and with the technology behind it. We, through using the technology and, by our use of it driving its further development, become ourselves vastly differently enabled. We are, thanks to digital, or 'intelligent', or virtual technology, capable not only of doing things we hadn't imagined before, we are also capable of *imagining* things we haven't imagined before, and we are giving ourselves the potentiality that comes with being able to do so.

And this same principle is one that we are now going to be able to apply not just to information but also to energy. So we can not yet quite imagine how exactly this will express itself, but what we can foresee is that the effect will be an opening up and *enabling* on a scale unlike any we've seen or experienced before.

But how does it *work?* How *does* information technology and energy technology converge, and what does that mean in practice?

There are three basic elements to this:

1) Energy *generation* or, more accurately, energy *sources:* where the energy comes from, the point at which it *enters* the system and how it does so.

2) Energy *consumption* or energy *'sinks':* where the energy is used and what for: where it therefore *leaves* the system.

3) Energy distribution or *logistics:* the layer that connects the two. We like to think of it, and call it, energy logistics (rather than 'distribution'), because as we have already seen, it is not a case any longer of getting energy from power stations to consumers, it's a case of orchestrating energy around the system, from all manner of points on the network to all manner of other points on the network, from wherever it has been 'sourced' to wherever it is being 'sunk', with many possible points of exchange and trade in-between.

The most popularly held belief is that the two critical elements in all this are a) energy generation/sourcing and b) energy consumption/sinking, and that it is therefore these two things we should concentrate on getting right, and address as a matter of utmost urgency the ways in which we use, waste, and could therefore save energy.

We realise, of course, that they are both important, but we think that it is in fact the third layer, energy *logistics* and energy trading, that is the key element which, if we get this right, can unlock *abundant* energy for us; and if we can do that then we don't even have to worry about how much we use and what for.

And that is why we will look at this middle layer, energy logistics and trading in the network, first and get to the other two afterwards:

V   NETWORK TERMINOLOGY

What exactly is the 'energy network'? Are we talking about one network or about several? Is this about what often gets referred to as 'smart grids', or about something else?

The terminology here can be a touch confusing and also misleading, because we often talk of electricity networks when in fact we mean grids, and we talk about grids when we mean networks, and we are beginning to introduce into an already complex equation a new way of thinking about energy, while keeping intact, more or less, an outdated infrastructure.

So we shall explain:

At the heart of what we may understand as an abundant energy future lies the concept or idea of a global energy network. This sometimes also gets referred to as the 'internet of things' or the 'internet of energy'. It's important to understand that this is not the internet, but something that works along a similar principle. The internet already exists and it's not a bad analogy, as from it we are totally familiar with the idea of transferring something – an email, a piece of information, a picture – from one place to another, freely and at lightning speed, at no or very little cost. The energy network allows us to do something very similar, but with energy.

The difference between 'smart grids', which indeed we are talking about on the one hand, and a 'global energy network', on the other, is not so much one of characteristics – they have all their essential facets in common – but mainly of scale. Because traditional electricity grids have grown up locally, regionally and nationally, smart grids tend to be conceived of and planned at a similar level. This is not wrong, it's just conservative. We want to take a broader perspective, and so we are extending the thought from local, regional and national smart grids to its logical ultimate implementation: a global energy network similar to the global internet.

So, having established what we mean by 'the network', we are ready to look at what happens in the network. And perhaps it won't come amiss to get to grips with this at the most fundamental level and start with the one thing that makes all of this possible:

VI   OUR FRIEND THE ELECTRON

When we talk about 'data', and about 'information', we strain to imagine what it is that actually happens with it in the network. We always see the result of data travelling through the network, because the computer goes 'ping' and we can read an email that's been sent to us, or we can watch a news item being streamed to our mobile while we're at the station waiting for the train, but it's well nigh impossible to visualise what actually happens in order for us to be able to do so.

That's until we make friends with the electron. Being a subatomic particle with a mass of a bit more than one two thousandth of a proton, the electron is very small indeed. If you were to place an average size grapefruit next to an average size pea, you would have, in very broad approximation, the relative sizes of a proton and an electron. Except that inside the atom they would be rather further away from each other. To get an idea of how far away, put your grapefruit in the palm of Nelson in Trafalgar Square in London. You would then find the pea whizzing around the M25, which, if you don't know London that well, is the suburban orbital motorway which encircles the city approximately 20 miles from the centre. Bearing in mind the pea size of our electron relative to the grapefruit nucleus in Nelson's hand, we then have an atom the size of Greater London. But an atom is hardly the size of London. An atom itself is so small that one million atoms, lined up next to each other, make up about the thickness of the page you're holding in your hand, if you are reading this book on paper. So you can see just how small an electron really is, it's as big as a pea inside an atom as big as London, but it takes a million atoms to make up the thickness of the page in a book.

Tiny as it is, the electron it is still negatively charged. In fact you may recall it has an elementary charge of -1. And that is why, in energy terms, the electron is our friend. Because being so small it can travel exceptionally fast, at nearly the speed of light. And having an electrical charge, even a tiny one, it will change the state of any atom it happens to travel to, because most atoms most of the time are balanced in terms of their electromagnetic charge, which means they are neutral. (If that weren't the case, you'd continually get small or large electric shocks when touching things. But as you know from experience, that only happens when you either touch something that has been deliberately electrified, such as a wire with a current running through it, or when you touch something that has got accidentally charged, such as a metallic surface that has been exposed to some friction, or the hand of somebody who's been moving about a lot in a synthetic garment.)

Because the normal state for most things when nothing is happening to them is neutral, each time an electron comes along with its negative charge it upsets things a little. The atom where the electron has arrived is now either negatively charged too, because the balance is out of kilter, or it, the atom, does something drastic, like expunge another of its electrons. It's the expunging that's the interesting bit, because the electrons that are already there really 'want' to be there. They don't 'want' to leave. (The electron, being a subatomic particle, does not, we know, have a 'will', we are trading metaphors here, obviously...) So there'll be a fair bit of jostling, before one of them goes, and that jostling is energy which registers as either heat or light or a magnetic field. And so although the dimensions are crazy, and there's an awful lot of nothing in atoms, the electrons, with their diminutive size, have a fantastically big impact on them. They are a bit like a courier. We are oversimplifying things to some considerable extent now, but like a courier, they can carry a message (information) or a log of wood (energy).

And the reason they can do either is because we have, about a hundred years after we started to make use of electricity as energy, found a way of using energy to symbolise information. We decreed that an electric charge should signify 'on' or 'yes' or '1', and that no electric charge should signify 'off' or 'no' or '0'. And we worked out that if you break things up into small enough units, you can codify any piece of information in precisely these two contrasting expressions: 'on' or 'off', which is the same as 'yes' or 'no', which is the same as '1' or '0'. (This, incidentally, covers only one technical aspect of digital encoding. In order to make information encodable, as it is today, mathematicians and then developers and inventors had to break with century old traditions and invent a whole new algebra, as well as programming languages and bit-based binary code itself. In the Western tradition, this goes right back to Francis Bacon,[LXXXII] who already talked about representing the alphabet in strings of numbers, but really was ushered in by Gottfried Leibniz,[LXXXIII] who came up with the foundation for the binary code system we use in computing today.[LXXXIV])

The 'digital' age was born, which is why to this day you see graphic designers illustrate all things to do with computers and information technology with cascading, travelling, floating or otherwise animated noughts and ones.

The achievement of using electricity to codify information has given us the ability to network information and deal with it, share it, distribute it, across the globe. But if we can 'symbolise' information using electricity – which we patently can – then we can also 'symbolise' electricity, using information. Now we no longer restrict ourselves to saying 'electricity can carry information', we can also say 'information can guide electricity'. Because it clearly can: we don't have to treat electricity as if it were a barrel of oil or a heap of coal. It isn't. It's an abstraction of a barrel of oil or a heap of coal, or anything else we like to use to 'generate' it. We can treat it as such, we can treat it as information.

(It may be of interest to note, by the way, that up until now it has not been possible to utilise individual electrons in information technology. As of now, we are always using bundles of electrons, but there is a lot of research being carried out in this field and we can expect significant advances in quantum computing over the coming years and decades, the impact of which may readily dwarf anything we've seen so far...[LXXXV])

VII   THE ENERGY NETWORK: AN 'INTERNET OF THINGS'

The internet of information already very much exists and is a full-on everyday reality in the majority of post-industrial societies. The global energy network is as yet mostly an idea. Just as the internet of information connects small local networks, so the global energy network will connect smaller, regional and local networks with each other to form one large, worldwide network. And like the internet, this will rely on some very heavy trunk traffic between countries and continents while at the same time being made up of innumerable branches, twigs and nodes of many a size and type.

In the internet of information we have nodes like Google, which handle billions of transactions of information every hour, and we have nodes like the laptop of your friend Florian who logs on once a fortnight to check his emails, of which there never are many, because apart from you he doesn't have many friends, and he likes to spend his time going for walks rather than writing emails to the few friends he has. Your smartphone is a node on the internet, as is the bank that has all your money. And they, within their own set-up, will have thousands if not tens of thousands of individual nodes also connected to their own network and to the big network that is the internet.

With energy networks it's just the same, and it doesn't matter whether we are looking at regional or national smart grids or at the yet-to-take-shape global energy network: it will consist of many millions and in time billions of nodes, ranging – and this is the new thing that we'll be explaining over the next few pages – from tiny LED lamps such as you may have in your garden, to giant power stations, such as you may have on the edge of your city.

What energy networks, big or small, regional, national or global, have in common is this:

1) TRAFFIC FLOWS IN ALL DIRECTIONS   Unlike a traditional electricity grid, in which power can only really cascade from the top to the bottom, from the power station via substations to the end user, in a genuine energy network the power can go from any point or node to any other. At its most radical and most evolved, a node can be anything: quite literally any *thing* that is or can be connected to the network by means of electricity, be it a hair dryer, a phone charger, a lamp, a fridge-freezer, or an electric car. This, it might seem, is only vaguely interesting, because all these devices would appear to use energy, not generate it. But that is purely a question of 'chicken and egg': these devices are in the main energy consuming, because up until now that's all they could be. In an 'intelligent' energy network, that is no longer the case: a device may as easily be energy providing, energy storing, energy exchanging or energy trading, as it is energy consuming. Each device becomes a dynamic participant in the network, much in the way that each computer in the internet of information does not 'consume' data, but rather handles it: receives some, sends some, processes and stores some.

Therefore, in an energy network, at its best, you can have any number of devices that can do any number of things. Your fridge-freezer, rather than simply drawing power come what may, can become capable of adjusting its energy requirement to the level of availability in the network and, for example, turn itself down or off for short periods when there is big demand.

Similarly, a charging device, such as a car battery, may realise that it doesn't really need the energy it has stored up for the next half hour or so and temporarily send it back into the network, because another device could use it. Also, there may of course be – and already are – any number of devices that are principally designed to generate or garner energy. At the moment we are thinking predominantly of solar foils or panels, and in fact these are of particular significance for reasons which we've already touched on and will expound along the way. But there is no law in physics or elsewhere that says that other types of devices – maybe as yet undreamt of and way beyond traditional mechanical ones like windmills or water turbines – could feed significant-enough quantities of energy back into the network.

And we do well to keep an open mind as to what is 'significant' in this context and what isn't. Many a drop fills an ocean, as we are all too aware in the reverse constellation: when it comes to consumer electronics or electric light bulbs, for example, we keep being told how even tiny quantities, replicated over and over again, make a big difference to our overall consumption. And this is certainly true. Equally true though is the fact that many devices being able to generate or harness tiny amounts of energy will also potentially make a big difference. We should probably not, at this stage, dismiss anything as 'insignificant'.

We shouldn't do so because the field opens up so widely and because the boundaries between energy 'generation' and energy 'consumption' get so blurred, but far more importantly because our understanding of what energy is and where it comes from and where it goes has changed. It doesn't in fact make sense to even think in terms of energy 'generation' and energy 'consumption' much longer. It makes more sense to think in terms of an energy system where energy enters at some points and leaves at other points. The processes are interchangeable and reversible, and the energy is not in that sense 'consumed' or 'used up', it is transferred from one node on the network to another.

2) ENERGY NODES BECOME THE 'CONSUMER PRODUCT'　　So energy, in this new model, is no longer something that is 'generated' or 'produced' and then 'consumed', it is something that is channelled or orchestrated. Because really the energy is already there anyway: 'At the moment, the energy is on the roof, where the sun happens to be shining, but I have a few sun foils on the roof, and so I can send you some of this energy straight away, it really is no problem at all.' I'm not 'making' the energy, not 'generating' and not 'producing' it, all I'm really doing is taking it from one place where it is in plentiful supply and sending it to some place else, where it may not be in plenty enough supply. And if a little more energy is required, I can just go out and buy a few more of these sun foils and also hook them up to the network, and a little more of the energy that's on the roof can now be shifted to elsewhere. It is quite that simple. In theory.

In practice it is a little more complicated, but not so much as to pose an insurmountable problem. The significant point is that when up until recently we thought mainly in terms of buying energy and thus treated energy as a consumer product, we will no longer be doing so now. We will be treating the device that either harnesses, or uses, or stores energy as the consumer product. And we are dealing with this orchestration of energy as a service provision which companies – most likely companies we have up until now regarded mainly as purveyors of energy – will be able to offer in a competitive market.

3) WE BECOME ENERGY TRADERS　　The above being the case, our role within the energy network changes. We will no longer be consumers who simply buy energy from an energy provider and pay a monthly or quarterly bill, but will instead become energy traders, who sometimes buy energy, sometimes sell it, and quite often might be doing both precisely at the same time.

This means that there will be extreme variations between the types and sizes of energy provision, energy consumption and energy trading that take place at any given time. If you happen to be part of a large energy conglomerate that owns several dozen power stations, you will be very big on selling your energy. If, on the other hand, you are an aluminium manufacturer, chances are you will consume vast quantities. If you are a farmer, you may use quite a bit of energy for your own needs at certain times but at other times have significant surplus on your hands that you can sell.

If, like the majority of people, you're simply a private person or a small business, you will, depending on where in the world you are and how you equip your flat, house or office, either be able to provide somewhat less energy than you use, or somewhat more, or about the same amount, subject largely to time of day and season. How much you pay and earn for your energy will be directly dependent on all the factors that make up your energy dealings: sometimes you will be using energy when nobody else wants it, so it will be extremely cheap – the 'happy hour' of energy trading – sometimes you will be using energy when it's in high demand, and you will have to pay a premium for this.

4) ENERGY IS TRADED IN AN OPEN AND CONTINUOUS MARKET
There will, as a result of all this, be new patterns and systems emerging as to how we pay and charge for energy. There is no reason not to imagine, for example, that similar to the way in which you don't pay your internet provider for time spent online or quantities of data transferred, but simply have a flat monthly usage fee that is set within certain parameters of data speed and 'reasonable' or 'fair' use, an energy service provider might offer you a deal whereby you can have a certain 'bandwidth' of energy for a flat monthly charge.

At the other end of the spectrum, it is equally conceivable that unlike today, where you are most likely charged perhaps a night time tariff and a daytime tariff for your electricity, you may be offered vastly varying prices throughout the day and week, which your domestic or business energy system will be configured to exploit to maximum effect and benefit to you. So your fridge would realise, for example, that half time of a football game when everybody turns on their kettle is a signally bad time to go into cooling mode. It can easily wait twenty minutes, without its temperature dropping enough to make any difference to the milk it's keeping fresh. This could be reinforced with pricing incentives: the market might register that on a hot summer's day there is big demand for cooling, and so air conditioning units all over town go into overdrive. Therefore the price for electricity goes up. The air conditioning unit in your annexe though in turn could be aware of this and confer with the motion sensor to find out whether you're actually working there this afternoon or not. Realising that nobody has been in, it could decide to lower its activity, allowing the temperature to increase slightly for that day, without anyone ever really noticing.

Having explained *why* we should network energy, we are of course going to explain also *how* this can be done. But before we do so we need to detour, briefly, and take a look at what it is that we use our energy for. Because it is certainly the case that we are saying, as one of the principal tenets of our thesis, that ultimately it will be possible to cover practically *all* our energy needs from sources so plentiful that the quantity of energy will no longer be an issue. If and when this happens, it will not be necessary to think in terms of saving energy any more. But it is also clear that this is not going to happen by clicking our fingers. We think it can and will happen sooner than most people are prepared to dream of right now, but even so there will be a period, of, we estimate, about 25 to 35 years, during which multiple approaches have to run side by side. One of these is networking energy, one is expanding greatly the number of installations for garnering energy from the sun and the weather, and a third one will be to make most sensible use of our energy too.

Making sensible use of our energy also requires something of a rethink, but this is very much underway. Architecture, engineering, government policy, they all work towards making buildings 'greener' and more 'energy efficient'. We will be looking at some of the concepts and also some of the related issues that this brings up. We will be asking, for example, to what extent it makes sense to talk in terms of energy efficiency, when perhaps energy integration may get us much further: making different technologies work hand in glove to the desired effect.

But here it is worth simply highlighting the point that today, about 40 % of all the energy we use goes towards getting the temperature right in our buildings. Heating and cooling the places we live, work and play in is a mammoth task and we spend *a lot* of energy on it. Much more than we need to: contemporary construction technology now makes it possible to build entire office blocks that require virtually no heating. Residential houses, too, can be built so as to use a fraction of what they did in the past. But it is obvious that this very large chunk of our energy demand is not going to go down easily: European cities, towns and villages are made up of buildings that date back as far as the 13th century, with architecture from any period between then and now, while many American cities are products of the early 20th century, and you find similar situations wherever you go. The world's housing stock is not going to upgrade to the latest technology in just a few years.

Still, we will want to, over time, migrate many millions of buildings away from fossil fuels to other technologies. Which ones will depend very much on the specifics of each case, there is no blanket solution. In some places it will be thermal solar energy that will work beautifully, in others it will be geothermal. In others still it may be a combination of approaches that creates the best balance between environmental concerns, heritage priorities and the demand for a steady, economical energy supply. And naturally, in many cases simply making buildings more energy efficient will go a very long way to meeting the short term challenge.

For many people, alarm bells start to ring when you say things like 'we will want to migrate' something or someone from one technology to another. Because that suggests massive master-plan style coercion. Which is neither what we have in mind, nor what will be necessary. Policy can and does create incentives, and there are clearly discussions to be had about what these policies should be, but we think that the attractiveness of the new solutions themselves will be the greatest incentive of all: the increasingly low cost of photovoltaics and solar fuels, the comfort and convenience of a networked home, the ability to feed energy back into the network when it is not needed and the chance to take advantage of competitive offers in an open market.

The design and user benefits of these new technologies are already pulling people towards them. As yet, this is happening mainly at the high end of the spectrum, but it will invariably cascade down to the middle and lower end too, as the installations and devices become cheaper and more familiar. So we are not advocating spending vast amounts of public money on advancing or preferring any particular technology. In fact the opposite: we are confident that given a level playing field, the new, networked approach, in combination with localised, location and site-specific harmonious thermal technology solutions, will win hands down.

And so two questions now urgently push themselves to the fore: if the energy network is such a great thing, why hasn't it been created before? And if now it can be done, then how?
Answering these questions is what we'll be doing next.

MAKING IT HAPPEN: A TURNKEY TECHNOLOGY

VI

If you were in mainland Europe at 22:09 in the night of 4th November 2006, sitting at home watching telly, at the cinema enjoying a film, in a bar having a drink, or on a tram heading home, chances are you may remember the Norwegian Pearl.[LXXXVI]

The Norwegian Pearl is a majestic and very large cruise ship that weighs 7,500 tons, is nearly 300 metres long and can accommodate four bodies shy of 2,400 passengers, plus 1,099 crew. It was built by the German Mayer shipyard in Papenburg and took its maiden voyage on 30th November 2006 from Rotterdam to Southampton, from whence it cruised on to Miami. Before doing so, it sailed into the consciousness of several million Europeans by interrupting, involuntarily, it has to be said, and unaware of it at the time, their evening's proceedings, in some cases for several hours.

On their way from the Mayer shipyard in Papenburg to the North Sea, vessels have to travel down the river Ems where they glide close by a place called Weener, by the looks of it a charming little town that sits directly on the border between Germany and the Netherlands. Here at Weener, there is a now notorious high voltage power line that crosses the river Ems and so connects two large, flat expanses of Northern Europe on the electricity grid in an east/west orientation. For a very large ship, like the Norwegian Pearl, this power line routinely gets shut down temporarily to allow the vessel to pass without danger. Normally, this procedure happens without hiccup and up until the Norwegian Pearl came along, hardly anybody even knew about it, let alone cared.

But the 4th of November 2006 was a windy night. And the many wind turbines in the area near the North Sea produced a lot of electricity. Normally, all that electricity gets channelled out of the region over two high voltage cables, but because of the Norwegian Pearl's passage to the sea, the Ems crossing was disconnected and so the power only had one line to travel through, elsewhere. All of a sudden, the grid had an unusual problem: on one side of the river Ems, there was way too much power, while on the other side, there wasn't enough. On both sides, the grid broke down. The result: large parts of Europe, stretching from Germany to France, via the Netherlands and Belgium, down to Italy and Spain in the south and to Austria and Croatia in the east, suffered blackouts. In Germany alone, some 100 trains got stuck, and France reported its most serious outage in 30 years, with roughly five million people affected.[LXXXVII]

Electricity grids are difficult to manage. They can cope with fluctuations in supply and demand, but only within fairly narrow parameters, and only when they're prepared. Which is why they spend a lot of time and effort researching consumer behaviour: to be able to predict and cater for it. And when an unexpected combination of factors causes a surge on the supply side, the grid is no happier than when there is a sudden surge in demand. So any system that hopes to handle very volatile behaviour both on the consumer *and* provider side has to be extremely dynamic indeed. And that's what has hitherto proved so difficult.

A GENIUS PLANET

A lot of work has been, and continues to be, done on power grids to make them more flexible and better able to cope. And indeed a lot of progress has been made. But the challenge is this: for an energy network to become truly effective, you have to be able to make each and every device – no matter how small, no matter how big, no matter how cheap or expensive, no matter what it is intended to do or what it actually does – *system capable.* In other words, everything that you are able to connect to the network, be it a chain of small fairy lights for your garden or a large wind farm, be it a set of solar foils on your roof or a car that's charging up its battery, has to form a dynamic node on the network, and this in turn means that it has to:

- Know what it is
- Know what it's for
- Act on this information and communicate with the rest of the network

Why? Because that's what makes the network as a whole dynamic. It's what allows the network to channel energy in all directions, and even out the many minor and major fluctuations. We have used the term 'orchestrate' a number of times now to describe what we want to do with energy, and this is exactly what becomes possible when each element on the network can actually be integrated into a system: it's what allows the network to understand how much energy is being fed into it and what to do with that energy, and it's what allows energy to be traded online and in real time amongst a multitude of participants, so that overall stability is always guaranteed. It's also what enables devices to act in solidarity with the network and adjust their demand if and when they can, which has crucial benefits for the system as a whole, and is also what allows your household to go for the best energy tariff and thus alleviate pressure when demand is high.

Making every device system capable is a tall order. We estimate that there are, as you may recall, some 500 billion of them, and the number is obviously growing. Integrating a few hundred power stations and large-scale electricity generating or consuming installations into a network and making them system capable is not so much of a problem. The investment, compared to the volume, is reasonable and the benefits will soon make themselves felt. The big challenge is doing so with the other few hundred billion devices, especially the very small and really old ones. It's a massive undertaking, and not one that can be imposed on people.

Imagine this in practical terms: your household, if it is an average European or American one, is likely to contain between 100 and 200 electrical devices. If somebody came along and said to you: you can have an extremely good deal on your energy, and you can be sure that all your energy comes from absolutely sustainable sources, and you will, by doing so, not only get very substantial user benefits, but you'll also do your serious bit towards saving the planet, chances are you'd be thrilled. But if you were then told that to this end you had to equip every single device in your home with some form of communication capability, and that this form of communication capability would obviously need its own power supply for it to work, your thrill would most likely sharply subside.

There are any number of issues that crop up the moment you set about making an existing infrastructure 'system capable'. But that is exactly what needs to happen, because *replacing* the existing infrastructure with a brand new one clearly isn't an option: you would not be able to rewire every house in every city, town and village, and at the same time replace every toaster, every washing machine, every fridge and every curler. Remember the hoo-hah when the European Union started phasing out candescent light bulbs? That would seem like a minor tut, compared to what you'd have to deal with here. Also, you need to allow for the process of conversion to happen gradually, and for the system to start working long before it is complete. In fact, you can't expect for it ever to be complete, because there will always be some people who for some reason or other will not want to have their house or their flat become part of a network, and there will be many people who will maybe want to give it a try first, with one or two devices, and see how it works before committing themselves to it on a grand scale. So flexibility, adaptability and scaleability of the system overall are all of the utmost importance. As are four factors that come into play at the basic level of local installation. These are the issues that are of primary interest to the homeowner, the consumer:

a)   Cost
b)   Ease of installation
c)   Power supply
d)   Secure communication

And here is where the turnkey technology that we now want to talk about comes in. Because getting one or two of these elements right has not proved all that difficult. What has eluded the scientific and engineering community so far is developing a system that meets all four of these challenges. Which is why we like to refer to this as the problem of the last yard: we have been, for a while now, so nearly there. But just not quite...

I   GENESIS OF AN INVENTION

When we introduced ourselves and explained why we were writing this book, we mentioned, a little in passing, that Ludger is an architect and an inventor and that one of his inventions is a key element to making energy networks a reality. This is now becoming quite important. Because as we go into explaining the technology in more detail, it will help us to bear in mind what our background to this is. None of us are politicians, none of us are members of any influential NGO, nor do we represent any particular sector of the energy industry. But we are talking from a point of considerable technological insight. And so we can say with some certainty that what we are proposing here is not a fanciful dream, but a theoretical as well as practical, technological pathway that we have confidence in because one of us has helped develop it.

As an architect and computer scientist, Ludger likes people, he likes ideas and he particularly also likes buildings. And he understands them. What he doesn't like is ideology and the rhetoric that goes with ideology. Nor does he like unused potential, and he particularly dislikes barrens of the imagination. Which is why, a few years ago, he became very fed up with the prevailing tone of the energy conversation. The tone was pessimistic, in some cases borderline apocalyptic, on occasion fatalistic. None of which helps matters get any better. What we needed, Ludger resolved, was a fresh optimism and a technology-inspired enthusiasm for the future. Flights of fancy grounded in application, was what he wanted, and pursued.

At the same time, Vera was writing her PhD on virtuality, examining what lies behind our use of media and networks, information and data. Their shared interest in what we have termed 'the potentiality of technology' is what brought Ludger and Vera together, and their concern for what was happening in the media on the subject of energy and climate change – the opinions that were being formed and the ways in which they were expressed, the approach that was taken and what that approach started to signify for us as a society – prompted them to look for a writer, Sebastian, who would help them develop their thinking on the matter and put it into words. And it wasn't entirely straightforward either, because while the three of us were working on the material, the thinking evolved quite a lot and so what you're reading today is very much an advanced and to some extent distilled intermediate result of an ongoing process.

But we can't tell this story without introducing one more character now: Wilfried Beck.[LXXXVIII] Wilfried started out in the early 1980s with an interest in computer games and soon began specialising in machines for gaming arcades. For this market, he developed the first PC-compatible chip and operating system. At the time, this was a big step forward for the games industry, since most game machines were then using their own, very limited and therefore limiting operating systems. Being able to run on PC platforms made them infinitely more versatile and a lot cheaper too. Wilfried then sold the company he'd set up for the purpose, and this allowed him to emigrate to the United States, where he started working in the area of 'smart homes', equipping East Coast villas with home automation systems.

By the time Wilfried and Ludger met, Wilfried had been looking for a business and development partner for some two years, while Ludger had been gaining a lot of experience in developing automation processes for large buildings. One of them was Microsoft's German Headquarters in the somewhat unfavourably named town of Unterschleissheim, just outside Munich, in the year 2000, which made it probably the first building to be equipped with this technology in Germany. Here, Ludger networked an entire large-scale office block over the internet, and the same system was being implemented by the same company at the European Central Bank in Frankfurt, the biggest building under construction in Germany at the time of writing.

The problem Wilfried Beck encountered in the course of his work was similar to that encountered by Ludger Hovestadt in his: systems electronics only become really affordable once you start producing them in very large quantities. This is the same throughout the electronics industry, which is why small, personal units, like cameras, calculators, MP3 players and mobile phones, once they catch on, become very cheap very quickly, whereas professional/industrial products are and remain expensive. Not only is it cheaper to produce units in large numbers than in small ones, but very importantly, the development costs, which are often immense, can be distributed over many more customers. So while it is complicated, difficult and expensive to make bespoke solutions that will, in the end, see some ten thousand units, incurring the same effort and expense for ten million units suddenly makes a lot of sense.

Ludger and Wilfried thus experienced, each in their own field, much the same difficulty: you don't get, in industry electronics, the kind of unit numbers that are necessary in order for prices to come down. Which means that 'intelligent' industrial infrastructures end up between 5 to 10 times as expensive as their 'standard' equivalents. And so in the past, equipping a house with network technology could cost anything from around $20,000 to $1.5m, but what you got for your money was not in actual fact all that impressive. The idea of the kind of 'intelligent' home that we see in films or that a Bill Gates [LXXXIX] was in a position to have built for himself is not in essence that new. As a concept it has been around – and fictionalised – for a good 20-30 years. Yet for the large part, and for the majority of people, it remained unattainable. Not only was the original installation cost prohibitive, but also the difficulty and subsequent expense of maintaining systems and keeping them up-to-date meant that, in all but the most exceptional circumstances, the dream of the 'intelligent' home remained just that, a dream.

But cost is only one aspect that hitherto presented an obstacle. Another is consumer behaviour. By and large, people don't do as they're told, and quite rightly so. Without even trying, they will find ways of flummoxing developers in the way they use technology. So while for a large industrial installation you can possibly drill people, train them, and impose processes on them which they will feel obliged to learn, because if they don't they'll get sacked, or at least reprimanded by their boss, at home you have no such stick to wave at them, and not much of a carrot either. The ridiculousness of the average VHS machine, back in the 1980s, was legend. People would rather stay at home and be there to press the 'record' button, than spend an hour trying to programme the thing, go out and come home to find that it had recorded a beginner's class in flower arranging instead of the champions league game. And have you ever read a user's manual, or met anyone who has? User's manuals are like end user licence agreements: a great deal of effort seems to go into making them as unintelligible as possible, and most of them succeed at remaining entirely ignored. At home, people expect 'plug & play': you buy a piece of equipment, you take it out of the box, you plug it in and turn it on. If it doesn't work immediately, and in an obvious way, you think of it as ill-conceived and either take it back to the store, or you restrict yourself to using its most basic functions, while complaining to your friends that it isn't doing what it's supposed to do, and vowing never to buy another one like it.

These two factors combined, cost and user-friendliness, suggest that any solution worthy of the description needs to fulfil, from the outset and quite apart from anything else, these two principal criteria:

- It has to present a *low threshold* investment, followed by low maintenance, and
- It has to be intuitive and simply work.

That was the challenge that Ludger and Wilfried faced, and that up until then nobody had really been able to live up to.

II   THE MISSING LINK

We are aware that there is a slight danger now, in citing Ludger and Wilfried's invention, of this coming across as a sales pitch. So it's perhaps worth reiterating at this juncture that Ludger has since sold the company, Aizo, that he helped set up to develop the chip. Aizo, in turn has since been sold, and although the chip is still made and marketed under the brand name digitalSTROM, Ludger is no longer involved with the product we're about to describe. And we're describing it here not because we want to sell it to you, but because we understand it and know how it works.

Ludger and Wilfried's approach was simple. To start with, they acknowledged that in order to achieve low unit costs you need to be able to produce high unit numbers. And this meant not treating every installation as unique and therefore trying to come up with a bespoke solution for each one, but instead taking advantage of the fact that a computer is in essence a *general machine* (or, as Alan Turing[XC] put it, when describing his precursor to today's computers, a 'universal machine').[XCI] What a computer chip is *capable* of doing is dependent on its own specifications. But what it *actually does* is dependent entirely on the software it runs. So the thing to do, Ludger and Wilfried resolved, was to move away from trying to build installations that can achieve a certain set of tasks that you believe are necessary or useful, and instead say to yourself: 'I don't care, for the moment, what may or may not be necessary or useful. I have no preconception of what an "intelligent" house is or does. What is of the essence is that the house is system-capable, because once everything in the house is part of, and therefore works as, a *system,* the system can then determine what everything does, which means that the user operating the system on the one hand, and the software developers designing the software for the system on the other, can make it do anything they want. And that may include things we haven't even imagined yet.'

This, as we've already seen, is exactly what has happened, and continues to happen, in personal computing and mobile telephony. The thing that you carry in your pocket is really just a small computer. What it *can do* depends on the kind of processor that's built into it and the amount of memory it has, but what it actually *does* is entirely a matter of the software you run on it. And that means, for any smartphone about, an operating system and applications. Hundreds of thousands of them. Not so long ago (as recently, perhaps, as the mid-to-late 1990s), nobody in their right mind would have imagined that there could be hundreds of thousands of uses for a mobile phone. And if they did, they certainly wouldn't have been able to name more than a couple of dozen. What has happened is that by making the technology available and putting it into the hands of users and developers, an entire world of possibilities has opened up that has employed the technology in the service of its users, in often completely new and previously unimagined ways.

Ludger and Wilfried decided to concentrate on this as their principal point of focus: to make the house *system capable* by turning every device in it into a little computer that can communicate. What this effectively meant was equipping every device with a processor: a computer chip.

This chip is the missing link: it's the component that brings together information technology and energy technology and therefore provides the basis for treating energy as a logistical system. Which is why we refer to it as a turnkey element: it unlocks the door to making the energy network a functioning reality. But as we've just reminded ourselves a short moment ago: the chip, the computer, is the unit that offers capability. What actually makes it perform all its functions is the software that runs on it. And here Ludger and Wilfried made another bold move: while the chip they've invented is a piece of patented hardware, the software that makes it work is an open source project. This means that all over the world anyone who wants to can become part of its development and build applications to run on the system. This, while relatively common in information technology, had not really been done in the energy sector before. It's what made Ludger and Wilfried's approach unique.

But Ludger and Wilfried's ambition extended beyond making a computer chip or initiating the creation of software solutions for the chip. The way they expressed this was thus: *"We see ourselves as a think tank and as a moderator, harnessing the collective intelligence available for the purpose of pursuing the vision of a sustained living environment that is also in keeping with (our) nature."*

Which is why they set up the digitalSTROM Alliance. This was a not-for-profit organisation, established at ETH Zürich (the Swiss Federal Institute of Technology), and its stated purpose was to develop the technology invented by Ludger and Wilfried to a global standard. Specifically, the alliance described as its remit the *"technological development, certification of products, definition of hardware and software standards as well as user interaction."* Membership was open to all interested parties, be they companies, associations, academic institutions or individuals.[XCII]

So Ludger is on the one hand the inventor of a specific technology, and on the other hand he was also working together with a whole range of other people on establishing a standard for that technology. And standard matters a great deal: today, in a world where people travel and products retail globally, technology without standards is unworkable. When you take your wireless mouse with you to Hong Kong, you expect, and you find to your satisfaction, that the AA batteries you buy there are exactly the same as the ones you buy at home, and they work, no matter which manufacturer they happen to be from. It's the same of course with your MP3 tunes and the protocol your web designer uses for your website. A technology standard is not a means of control, it's a way of ensuring compatibility. And so, being a pioneer in this field, digitalSTROM obviously sees itself as well placed to play a leading role in defining an industry standard.

III   HOW IT WORKS

The basic set-up of an energy-networked site, whether this be a flat, a studio, an atelier, an office or any other small to mid-sized premises, is essentially this:

THE SERVER   You need a small server (a mini-computer) that runs a piece of operating software which controls all the other components of the system and channels and interprets the communication that takes place between individual devices within the system (for example 'TV to telephone' or 'blinds to light switch') as well as outside the system (for example 'fridge-freezer to network'). Like an internet router or a fuse box, it provides a central access point to your network, and most of the time you can forget it's even there.

THE OPERATING SYSTEM   Any computer, big or small, needs an operating system that tells it what to do. For personal computing the most widely used operating systems are Windows for PC, Apple OS for Apple Macs and the open source Linux, which you can run on either. Similarly, your smartphone will run an operating system, such as Android or the iPhone OS. The operating system is the signally most important piece of software on any computer, whether it's an old desk top machine, a laptop, a tablet or a smartphone, or indeed a server, because it determines how the computer works. Everything else, all the applications, are tailored to and determined by the operating system.

APPLICATIONS   Apart from the operating system that controls the server, you also need some applications that can make the system do specific things or allow you to carry out tasks, or make the system carry them out for you. For example, you may like the idea of your house having a number of presets for different occasions. So a 'holiday' preset might make it turn down the heating and air conditioning just so as to keep the house at a minimum or maximum temperature to avoid any damage, while switching the lights on and off at suitably random intervals to give the impression of somebody being at home. Or a 'relax' preset may put the temperature at a comfortable 22.5 degrees Celsius, dim the lights a bit but leave the reading lamp on and put on some classical music for you. There are really no limits to what kinds of applications may become available for a system like this, and as has happened with applications for smartphones, we would expect many thousands, perhaps hundreds of thousands of uses that individual applications will be developed for, possibly in a very short time.

THE CHIP   The fourth component that the system needs is a number of microprocessors or chips, installed in the various devices around the house, so as to allow each one to:

- Have their own unique identity (much as every computer on the internet has its own unique IP address, so every device on the energy network needs to have its own identification number that distinguishes it from every other device).
- Know what and where it is and what its energy requirement or provision is going to be at any given time. (A simple solar foil, for example, would never require energy from the network, but would at varying times of the day or night be able to feed very varying amounts of electricity into the network, whereas an electric car, by contrast, may at times be charging up with power from the network, while at other times being able for a while to feed power back into it so as to help bridge a temporary shortage.)
- Communicate all this with the other devices in the system, with the server, and through the server with the network, in real time, reliably, and without human input.

Of these four components, one – the server – is unproblematic, because it's a simple small box with some basic computing capability that can sit anywhere in the house. Two – the operating system and applications – are software components which are continuously being developed over time by a large number of people all over the world, and as we know, there is hardly anything you can think of that you cannot write an application for.

The crucial component and clinching factor therefore is the chip. This is where the difficulty has hitherto lain. Because for it to be of any practical use, the chip, much as we have seen with the system overall, has to meet these four principal criteria:

1) COST   The amount of money you as the user have to spend in order to become part of an energy network must not be prohibitive. In fact, it has to be so low that any benefits you may gain from being connected very quickly start to pay for your outlay. So first of all, the chip has to be *cheap*. Buying several dozen, maybe as many as one or two hundred of them (which you may recall is roughly the number of electrical devices

in a European or North American household) must not constitute an insurmountable problem for the average householder. In concrete terms, we are talking about a chip that costs somewhere between about 50 cents and two or three euros or dollars a piece.

2) EASE OF INSTALLATION   Tied in directly with cost is ease of installation: it just has to work. Do you really want to call around an engineer every time you alter your set-up? Do you want to call around an engineer in the first place? Ideally: no. Ideally, you buy it in the shop or order it online and take it out of the box and connect it, and there it is: plug & play.

Apart from that though, any device equipped with the chip installed has to also work just the way it did without the chip. In other words, it has to basically perform like any ordinary device, doing whatever the ordinary user would expect from it, and then perform any additional functions that it may have as a result of now being part of a system on top. So if, for example, you buy a new bedside alarm clock, what you expect to happen is that you plug it in and it shows you the time. You then may have to press a few buttons to set the alarm, and once that's done you know you'll wake up in the morning. This still needs to be the case even if the alarm clock now becomes part of your 'smart home' system. You should not have to learn a whole new set of buttons just because your alarm clock can now talk to the central heating and tell it to turn itself on half an hour before the alarm goes off, for example.

3) POWER SUPPLY   The chip has to be able to use the existing wiring for its own power. This means it has to be a 'high voltage' chip. 'High voltage' is a relative term and in this context it means that it can feed off an existing 110V or 230V power supply. A lot of chips run on voltages of, say, 6V or 12V, which is why the corresponding devices routinely come with a power adaptor attached to them. But neither do you want dozens of additional power adaptors lying around your house, nor do you want cables, running from these to your dozens of chips. Nor do you want to be buying mountains of batteries and then keep going around your house all year long replacing them. So this is what's meant by 'high voltage' here (as opposed to the kind of high voltage cables you see running across the countryside with charges of several thousand volts).

4) SECURE COMMUNICATION   If all your devices are going to communicate with the network, then they have to do so in a way that keeps the data that they communicate secure. Because if the communication could be intercepted by an outsider, that person would instantly be able to gather a whole raft of information about the geography of your house and about your usage pattern and movements, which would allow them to sabotage or deliberately breach your security and that of your loved ones. Unsurprisingly, then, this is also the touchiest of the four subjects.

There are, on the surface, three available options for the devices' communication:
- Extra wiring – connect them with new cables
- Radio frequency – use wireless communication
- PLC (Power Line Communication) – 'piggy back' on the existing electrical wiring

Obviously, we need to examine them briefly in turn:

EXTRA WIRING   In technical terms this is called a bus system with its own cabling, and it barely qualifies as an option. No more do you want extra cables and adaptors cluttering up your home to allow your devices to communicate, than to supply power to the chips. Imagine running two hundred new cables to every lamp, toaster, TV, radio, electric shaver and phone charger: you'd go mad and you wouldn't be able to move for wires. And you'd quite likely incur an issue with electrosmog, not as a health hazard, but as interference with other communication devices in your household, and indeed with the communication of the devices you are connecting. For these reasons, we can forget about extra wiring.

RADIO FREQUENCY   This is a system that uses dedicated radio frequencies to send signals between electric devices. It doesn't, very obviously, pose the problem of physical cables cluttering up your space and it's easy enough to install.

It does, however, present two particular problems that are difficult to overcome: firstly, radio frequency uses a very low voltage whereas the entire existing infrastructure, as we've seen, uses a high voltage (normally either 110V or 230V). In order for radio frequency to work, these high voltages have to be converted to very low voltages, a process which in itself takes a lot of energy, as anybody who has ever noticed the power adaptor of their laptop, for example, getting hot over time will know.

The second problem with radio frequency communication is that it's difficult to make secure. If you let a large number of devices send and receive radio signals, then chances are that you'll get, at one point or another, radio interference or distorted signals, which means a breakdown of communication and therefore a cessation, even if only temporary, of the system as a whole. You do not want that to happen: imagine sitting in your car in the morning, with the kids ready for school, and the garage door refusing to open for no apparent reason...

More troubling, perhaps, your radio signals may be intercepted or deliberately tampered with by illegitimate or malign users. This may not be of devastating consequence when you're talking about the light in your hallway. But what about your alarm system or indeed the garage door? With our wireless internet connections we go to great lengths to protect our networks from viruses and to encrypt sensitive data. Now imagine instead of two or three laptops and a handful of smartphones per household, having two hundred pieces of equipment talking to each other.

POWER LINE COMMUNICATION   Power line communication, or PLC, uses the existing wire that feeds power to the device to also send the chip's communication signal.

There have, until now, been two basic problems with this. The first one is that previously existing technology uses a lot of energy which has the effect that the chip gets hot and therefore needs cooling. If it needs cooling it can't be built into the wall or sit in a small enclosed casing, as it needs air to circulate around it so that it won't overheat. The second, more fundamental issue is that these systems try to implement a very subtle, fine-tuned frequency (used for the purpose of communication) on top of a very robust high voltage frequency (used for the purpose of transporting power). Thus, the two frequencies can end up 'fighting' each other, causing communication breakdown and malfunction.

IV   THE DIGITALSTROM CHIP

Of the four factors we've just looked at – cost, ease of installation, power supply and secure communication – the biggest technological challenge is number four: secure communication.

For all the reasons cited above, Ludger and Wilfried decided that what they were after was a 'high voltage PLC' chip that could be mass produced at a low unit cost and that required no installation expertise at all. Importantly, it also had to overcome all the problems traditionally associated with power line communication. And that's why they invented the digitalSTROM chip. ('Strom' is electricity in German in this context; the name therefore literally means 'digital electricity'.)

So how does this chip overcome the problems traditionally associated with microprocessors that use similar technology: how does it make power line communication work without the need for cooling and without frequency clashes?

Put simply, instead of 'fighting' the electricity that's present in the wire, it's 'surfing' on it. And this is possible because of a characteristic in electric currents that hardly anybody who hasn't studied the matter is aware of, but that comes in very handy in this instance:

Conventional electricity standards for alternating current (you may recall that AC is the type of electricity used almost universally for domestic and light industrial purposes) allow for extremely short cut-outs to occur in the electrical current feeding a device, without interrupting its operation. In fact, such minuscule 'gaps' occur regularly, every time the current changes from + to - which ordinarily happens 50 or 60 times per second, depending on where in the world you are and which frequency your electricity network uses, 50 Hz or 60 Hz. So if you are feeding a light bulb, for example, the power running through the wire to the light bulb is in fact 'off' by up to 20 milliseconds at a time without the light bulb visibly going off. This means that if you send your low voltage current (that you need to let the bulb communicate with your server) along the cable in that 'gap' when the high voltage current is 'off', you can get your devices to talk to each other without any other part of the system being affected or interfered with, and without needing any additional energy to convert the voltage. It's like riding the wave, but not on the crest nor in the trough, but right in the middle. This, then, is what the digitalSTROM chip combines on itself:

- A 110V/230V adaptor
- A unique identity (akin to a computer's IP address on the internet)
- A processor
- Data storage
- A power line modem
- Plus some 60 other functions which are potentially useful for the device, such as a driver for energy saving light bulbs or LED lamps, dimmer switch functionality, touch and motion sensor inlets, network frequency gauge and electricity meter.

All of this requires a negligible 0.1 W of power (an energy-saving light bulb, at 10W, uses about a hundred times as much). So the chip itself makes almost no demand on the network, and because it is so versatile, it is suitable for any type of device, new or old. Manufacturers can pre-install it in anything from a small light bulb to a big TV, and for existing devices there's a simple attachment that you plug it into before it goes into the power socket in the wall. (If, instead, you want an electrician to come around and do all your light switches and built-in machinery for you, that's possible too.) So anything that uses electricity can be quite easily and cheaply equipped with the chip, and the moment that has happened, the device becomes system capable and can therefore become part of a network, whenever the user wants it to go 'live'.

This particular point may well be the most remarkable, and at the same time also the most practically useful thing about the chip that Ludger and Wilfried have invented: it *just plugs in and works.* This is literally the case: you can take a digitalSTROM chip out of the box, plug it into any ordinary socket, plug any ordinary device into it and there you have it: the device now communicates directly with the system's server. There is nothing else you need to do.

What this also means is that you don't have to plan ahead. Imagine your home is an average two or three-bedroom flat and you think that going 'smart' with it would be a smashing idea. But the economic waters are a bit choppy and you don't know that you can invest in a complete overhaul of your home. This type of technology doesn't expect you to. You can start with a minimal set-up consisting of the server and one or two devices. If you like the way it works, you can add more devices to the system, as and when you see fit. Whether this then entails retro-installing them, by getting the small adaptor plug, or whether it simply means buying devices which have the chip pre-installed (as over time more and more devices will have), is entirely up to you.

How, though, can we be so certain that more and more devices will come with the digitalSTROM chip installed? The answer of course is, we can't. What we *do* know is that the digitalSTROM chip will be one of many different chips that will be available, and that therefore more and more devices will come with either this or another, similar, chip installed. And what we also know is that this one is so small — about the size of an ant — and so versatile that it really does fit into any plug or suit any device. Being so small, it uses hardly any power for itself, which means it doesn't get hot, and because it doesn't get hot it doesn't break down or cause fires. Like any computer chip, it will become cheaper as more units are made. And because it mimics existing behaviours, you don't have to learn a whole new way of doing things, you can just carry on doing what you've been doing the way you've always done, and gradually, as you get familiar with it, start using the functionalities that the chip offers.

Again: we are not suggesting that the digitalSTROM chip is the solution to all our energy problems, nor are we suggesting you rush out and buy your starter kit now. What we are suggesting that this is a good example of a working technology of which we can be certain in so far as we have amongst us the person who's invented it, tested it and is using it, even as we write this sentence. It's us saying: we are not looking at pie in the sky, we know that this can be done.

V   APPLICATIONS

We've noted a couple of times that information technology and energy technology are now coming together, and that this is precisely what an invention like the digitalSTROM chip makes possible, wherein lies its true significance.

A chip is at once one of the most useful and the most useless thing imaginable. On its own it does absolutely nothing. It could sit there, being amazing, and nobody would ever know about it. Without software a computer is pointless. But put some program, some code, some algorithm to it, and untold things become possible.

We've also already talked about operating systems and applications. As anybody who has ever used more than one of the major operating systems will know, there are vast differences in how user-friendly, how intuitive, how stable, even just how 'elegant' they are, and there are quite substantially different ways of thinking at work in their genesis and evolution. But the baseline principle remains the same, whether yours is a PC or a Mac, an Android, a BlackBerry or an iPhone. So if we cite the iPhone as a fine example, it's not because it's unique, nor because we mean to endorse it, but because it started the whole smartphone thing off and its statistics are particularly impressive: the iPhone App Store opened in July 2008, offering some 500 applications. One year later, by July 2009, there were 65,000 applications designed for the iPhone, which had been downloaded 1,500 million times. By July 2010, there were in excess of 250,000 applications with over 5 billion downloads. At the time we are revising this chapter in October 2016, this has risen to more than 2 million applications with over 130 billion downloads, and you can assume by the time you will first have a chance to read this book, the numbers will have risen again.[XCIII]

In information technology, there are some principles at work which enable radical developments. One of these is Moore's Law, which we've already mentioned, another principle at work is what one might call cluster clout: the potency of people coming together and working, independent of each other but jointly, on the development of a project or the solution to a problem.

Which is why Ludger and Wilfried decided to keep the software development for their digitalSTROM chip open source. It means that instead of belonging to a company or a corporation, the software belongs to everybody. And this in turn means that many more brains will be working on it, and many more people are bringing their ideas and their thinking to it – thus being able to offer their solutions to problems and challenges – than would be the case with proprietary software. In this way, important issues such as data integrity, security and confidentiality are being negotiated and worked out on a broad platform: they are not being determined by some central, self-interested body, but by the community of people who see the benefit in the technology. Ludger and Wilfried did not set out to build a system or a solution. What they wanted was true system capability and the potential for any number of solutions. They wanted to marry the imagination of information technology with energy provision and introduce into the home what we've seen in computing and mobile telephony: applications.

When the first mobile phones came to market in the late 1980s, they famously had the size of a brick, were carried around in briefcases by important people and those thinking themselves such, and all they could do was make and receive voice calls. They actually *were* portable telephones. When the smartphone was invented, in the late 1990s, it was able to do maybe a couple of dozen things. It could make and receive phone calls, it could wake you up and double up as a pocket calculator. It most likely had a camera, and of course you could text. We say 'of course', but it's worth remembering that 'texting' did not exist for the consumer before 1993. The first SMS text message was sent by a then 22-year old British test engineer called Neil Papworth [XCIV] from a personal computer to an Orbitel 901 handset on 3rd December 1992.[XCV] In 1995, the average number of text messages sent by customers every month was 0.4, or about one text per person every 2 months. The average smartphone today sends and receives hundreds of texts a month and does almost anything we can imagine, and quite a few things we until very recently couldn't, apart from putting the proverbial kettle on, though it very easily could also do that: all it needs is a small application and for your kettle to be connected to the network via something like a digitalSTROM chip...

This, really, is what we're getting at: the power of the imagination. Back in 1992 – and we remember that year very well – very few people could really *imagine* what it would be like to have an app for your phone. The fact that today we book train tickets, play games, read newspapers and watch videos, look up dictionaries and find out what that tune it is they're playing on the radio, all using a gadget that most conveniently sits in our palm, was all largely unforeseen. The reason it came about is because it *could* come about, and the reason it could come about is because the technology platform for it existed. In other words, nobody sat down and said: we really need to find a way in which we can make it possible for people to use their handsets to look up the menu of their local seafood restaurant and order a takeaway meal, because not being able to do so is a real problem and we have to find a solution for it. Much rather, somebody saw what was happening with smartphones and thought: wouldn't it be cool if I could do this. And then went ahead and created the app for it.

This same principle is now being applied to the home. And here it is exactly the same. We have no idea what kind of apps people will want or need or just feel like having for the home. So we don't want to venture very far into 'future scenarios': they tend to sound wrong and turn out to be so. What we do know though is that much as it is apps that make your smartphone do what it does, it will be apps that will orchestrate your home and the devices in it. And here, too, the possibilities are virtually endless.

VI   NETWORK SOLIDARITY

Right at the beginning of this section we touched upon one particular problem of our current system for the provision of electricity that makes it difficult to manage: you have to feed exactly as much power into it as you take out of it. We've seen what happens when demand outstrips supply or vice versa, the example of the Norwegian Pearl illustrates this in a dramatic fashion: the grid breaks down. It is difficult enough as it is, with a limited number of power stations and power generators, to regulate supply, and it's almost impossible to regulate demand. All power providers can really do is study consumer patterns and be prepared. As indeed they mostly are, if at the cost of vastly excess generating capacity that only kicks in at peak times. And with more and more different types of electricity generating installations of widely varying sizes coming online, the challenge of maintaining this balance of input and output becomes ever more acute.

Today's power stations in the developed world are capable of generating about twice as much electricity as is normally required. When Prince William married Kate Middleton on 29th April 2011, Great Britain's national grid experienced a surge in demand for power of 2.4 gigawatt, approximately the output of a nuclear power station, at 12:40 local time.^{XCVI} Why? Because at that time the couple had reached Buckingham Palace, and so about a million television viewers got up and put their kettles on.

In Britain, during a normal week in October 2016, electricity demand ranges from a low of 22 gigawatt between about Sunday midnight and 4am Monday morning to a high of about 42 gigawatt every weekday between about 8am and 6pm, with normally a small dip around the early afternoon.^{XCVII} While the absolute figures will vary from country to country, the relative figures are replicated almost anywhere you go in the developed world. At peak times, demand is nearly twice that of trough periods. Once you then factor in exceptional events (for the Royal Wedding above you may read almost any significant occasion that causes large numbers of people to do the same thing simultaneously, such as half time at a big football game) and you have a costly and complex task of responding to consumer behaviour on your hands.

During the Beijing Olympic Games in 2008, other parts of China had to reduce or even cease energy intensive activities, such as industrial production, to prevent power shortages from affecting the games.^{XCVIII} The cost of catering for these vast fluctuations, both in monetary but also in environmental and structural terms, is immense and runs into hundreds of billions of dollars every year.

So if demand could manage *itself,* then we could find that our energy generating requirement is in fact a lot lower than we think it is. In Great Britain alone, if the differential between peak and trough were two thirds instead of nearly double, say 14 or 15 gigawatt instead of 20 gigawatt, then that would mean giving everybody the same amount of energy as today, but about three of Britain's nuclear power stations could stop operating immediately, with nobody noticing the difference.

How can this be done?

Energy demand on the grid at any given time is measured by the frequency of alternating current. In Europe, this is 50 Hz (Hertz), in North America and parts of Asia it is 60 Hz. (These are domestic frequencies: European railways, for example, run on 16.7 Hz.)

Now, if the frequency falls by 0.2 Hz to 49.8 Hz instead of 50 Hz (or 59.8 Hz instead of 60 Hz, respectively), this signals that a lot of pressure is put on the grid by users, and the power stations have to work correspondingly harder to generate electricity and bring the frequency back up to normal. Just as cycling up a mountain makes your legs work harder and the wheels of your bike turn more slowly than cycling on a level plane, because more power is required, so the generators of a power station turn more slowly and have to work harder to cope with the extra demand, as a result using more of the energy that is put into them to make their turbines turn.

By contrast, if the frequency goes up to, say, 50.2 Hz (60.2 Hz), then this is a sign that the reverse is the case: there is less demand from users, and the power generators have to ease off, like a cyclist whizzing down the mountain who may have to put on the brakes a bit so as not to race out of control.

If, by adjusting their output, the generators don't manage to regulate the frequency to within a margin of minus or plus 1 Hz (49 Hz minimum, 51 Hz maximum or 59 Hz / 61 Hz respectively), the system has to be switched off, as it would otherwise break down with catastrophic damage that could take a very long time, and a lot of money, to repair.

What complicates the situation is the fact that most grids are fed by different types of power stations which vary greatly in the degree to which their power output can actually be adjusted. Belgium, for example, by 2016 generated some 55 % of its electricity in nuclear power stations,[XCIX] but until recently it used to be as much as 60 %, and since it is difficult, slow and expensive to turn off a nuclear reactor, at night, when most Belgians are asleep in their beds, their country used to have a lot of excess energy, some of which it burned off in its famous blanket motorway lighting, which, apart from the Great Wall of China, used to be one of only two man-made structures to be visible from space.[C]

In electricity generation, the minimum amount of power that is required to keep everything going when there is least demand (usually between three and four in the morning) is known as the 'base load'. This tends to be covered by power generators which are inherently difficult to adjust, principally continuous flow hydropower stations, brown coal, and nuclear power stations, all of which either do or are supposed to provide comparatively cheap electricity.

The base load coverage is complemented by a system of top up stations which are flexible to varying degrees. Gas power stations, for example, present relatively low investment costs and are very easy to adjust, but they are comparatively inefficient and therefore expensive to run, so they are used only when necessary. In mountainous regions, pump storage stations come in handy: when there is too much energy, they are used to pump water up into a reservoir or lake at a higher level, and then when the power is needed, the water is released through a set of turbines, restoring the energy to the grid. The amount of energy that is lost in the cycle amounts to about 20 %, which makes the method acceptably efficient.

The large fluctuations between base load and peak demand account for the fact that the European Union in 2012 had some 947,000 megawatt power generating capacity in order to cover an average requirement of about 316,000 megawatt.[CI]

A much more even demand can be achieved if non-essential power consuming devices, such as water heaters, air conditioning units or heat exchangers, realise for themselves when there is peak demand from other, more essential, devices, and are able therefore to briefly, unnoticeably, switch themselves off.

Certainly, if we are going to have energy abundance in the longer term, this may not really matter any more, but in the shorter to medium term, it matters a great deal. Cutting the generating requirement in half may be a touch ambitious, but 'intelligent' energy networks can realistically reduce it by about a quarter to a third, without anybody suffering a supply crisis. So for as long as there are large numbers of fuel-dependent, emissions-intensive power stations in operation, technology that enables network solidarity can play an important role by rendering up to about a third of power generating capacity obsolete.

It does this by measuring network frequency on a ongoing basis. Without you noticing or having to do anything about it, the chip itself can determine what the current demand is on the network, and instruct the device which it controls to behave accordingly. So when the network is going through a period of high demand, such as when everybody's getting up in the morning, putting their toaster and kettle on and running their shower, devices that aren't needed can lean back and effectively say: 'you go first'.

We've seen how a sudden overcapacity is no healthier for the network than straining under excessive demand. With this type of system, when it realises that the network frequency is rising perilously high (meaning that there is not enough demand), it can start taking excess power off it, by, for example, charging up a storage battery or running a washer dryer that has been on standby, or charging the car that's parked outside.

The more volatile and unpredictable energy sources are, the more important such flexibility becomes. Say you want to cover motorways with photovoltaic solar cells to generate electricity. This is entirely possible and makes excellent use of an already prepared, already flat, already covered part of the landscape. You don't make a stretch of countryside any uglier by putting a layer of semiconductors over the tarmac: turning roads into solar power plant makes a lot of sense. But, lying as it does across a long stretch of open countryside, any road will permanently be subject to fluctuating weather. Not just weather that changes over the course of a day or two, but weather phenomena that can change within seconds. You may start off with a pleasant day: the sun heats up the meadows, hills and forests; clouds start to form and for a few hours the sun disappears behind the clouds. As the day wears on, a bit of wind comes up and blows the clouds away: the sun blazes down onto the solar panels with full force again. Our motorway, having gradually clouded over during the day, at this point gets a sudden burst of energy that has to be dealt with immediately. If the cars charging up in the villages along the motorway, and the cold stores of the supermarkets in the towns, and the air conditioning units of the office blocks in the cities all realise *by themselves* that there is a power surge and take that power off the system, then not only does the energy not get lost or wasted, it also makes the overall energy provision stable, reliable and secure.

Networking energy, then, not only means being able to access many more sources of energy, especially many more *infinite* sources of energy, such as solar, wind, tidal, wave and hydropower, it also means being able to reduce the amount of energy generating capacity that is required to keep everything going, because network users can behave 'in solidarity' with the network, which in turn can result in very substantial cost savings for both the user and the energy provider.

VII   TRADING ELECTRICITY

Say you were based in Europe and ran a very big electricity-powered machine, something on the scale of a Large Hadron Collider, for example, which averages around 120 MW – the approximate equivalent of a small town – then Sunday 4th October 2009 between 2am and 3am would have been a good time to really turn it up and run it at full pelt: instead of paying somewhere between €2,000 and €5,000 for the hour, as you normally would, you'd have earned yourself the tidy sum of €60,000. Because on that day, for that one hour, the spot price for electricity at the European Energy Exchange (EEX) fell to *minus* €500 per megawatt hour.[CII] Indeed: you were being paid for taking power off the grid. It didn't last very long and was due to an unusual combination of mild temperatures and strong wind coinciding on a time of night when most people are curled up in bed, but the event was by no means unique, and it illustrates how precarious the industry has become.

Since the beginning of the 20th century, the energy economy has grown very substantially. Over a long time it did so at a rate of 20 % per year, similar to the growth rates we still see in the internet and mobile telephony industries today. It's been a robust sector, that has survived crises and wars and has, because of its unparalleled significance, for the most part been the focus of many and diverse political and economic interests. Both politics and economics are, we all know, inseparably tied up with energy. And for very good reasons.

One of the advantages of a centrist, pan-regional supply structure is that big power stations are more economical to run than small ones. Having said that, regional interests and priorities are brought into play by local governments and authorities who have legislative powers over water and highway rights. So what has emerged over the years are relatively autonomous supply regions which nevertheless form grid unions across large areas, sometimes across countries, to help prevent power cuts resulting from under or oversupply.

This has created an energy topology which is tailored to big power plant. For about 50 years or so, this topology has provided a degree of stability and security for Europe, but since the early 1990s the established system has found itself increasingly under pressure. Today, the principles of networked technology, as well as economic considerations, are beginning to take priority over the need to simply safeguard a steady supply. So in 1996, a EU directive separated energy provision, distribution and trading out from each other, thus creating a liberalisation of the energy market by preventing cross-financing in the industry, which was known to hamper independence and competition.

The European Energy Exchange was set up in 2002 as a direct result of this liberalisation. Here, futures (energy prices in the long term) are contracted up to six years in advance, and short term auctions determine the price of energy for the next few days, or, as in the case above, on the spot market, over the next few hours. Ordinarily, the prices for a megawatt hour range, at the time of writing, from about €20 to €60, but, quite apart from the fairly unusual minus €500 cited earlier, the market has also seen figures as high as €2,000.[CIII]

So everybody – energy producers, traders and users – faces the relatively novel difficulty of having to reconcile fluctuating prices with fluctuating demand. And while traders may well know the price they are going to pay for energy, they can only hazard more or less accurate guesses as to how much energy their customers are going to want, mainly by looking at statistics of previous use. So they often find themselves buying too much electricity in advance, to stave off sudden shortages, or, if they fail to cover for shortages anyway, they pay way over the odds for last minute extra supplies.

This problem area too can be addressed and put to both the end user's and the energy provider's advantage if electrical devices are capable of being part of a system. Because full system capability allows energy providers to transmit real-time information about current energy prices, which in turn makes it possible for the devices themselves to plan their energy consumption ahead, or respond quickly to favourable market conditions.

Many types of appliances and devices, such as heat pumps, hot water boilers, car batteries, fridges, freezers, air conditioning units, even washing machines can be quite flexible in determining when exactly they use their energy. If you have your washer loaded with laundry, it may make no difference to you whether the actual washing happens right now or in two hours' time, because you may be at work, or out shopping, for the rest of the day anyway. A fridge-freezer can easily go on reduced mode or even turn itself off for a short while without risking any dangerous or hazardous rise in temperature. In fact, a modern, well-maintained freezer can delay or anticipate cooling by several hours without ever affecting the quality of the food that's stored inside. Similarly, an air conditioned office building or residential block may be in a position to do a lot of its cooling at times when there is plentiful cheap energy available, and step back for short periods to allow for short-term pressures on demand. We have seen how some 40 % of our energy is used for houses (domestic and small business properties), and that a large portion of this in turn goes on heating and cooling. With an intelligent system it is no big problem to use excess energy when it is cheap and store it for later use, when it is expensive. Depending on the climate zone you're in and what you use much of your energy for, it may make sense to install thermo-solar panels that do not generate electricity, but heat up water instead and then use a heat pump to store the energy that you get during hot periods in the ground to be used later in the day or at night, when it's cold, for example.

The interesting thing about all of this is that there is no central planning involved, nor is any human interpretation or action required. Energy consumers and energy providers stand in constant dialogue with each other, without even noticing it. And the communication helps both the consumer, because they get the best possible prices, and the provider, because their supply peaks and troths are being levelled out.

One important detail to note is that this levelling simply requires a communication broadcast from the provider, meaning that the provider just has to have a channel of communication open towards the consumer. It is not necessary for the provider to know the identity or even the nature of individual devices or installations that use the energy. So the consumer's privacy is protected, while the system as a whole is still capable of stable and adaptable behaviour that is responsive at a level way beyond what would be possible on the basis purely of usage statistics.

None of this is possible without system 'intelligence'. There are currently big investments being undertaken in an attempt at making old electricity grids more capable of responding to fluctuating demand. But they're wholly inadequate when pitched against the level of volatility we can not only expect but predict. In Switzerland, as we write this, people are building hydrodams at 2 billion Swiss Francs a piece (a bit over USD 2 billion at current exchange rates) for the purpose of using them as energy buffers. And of course, that's one way of doing things, and it's not even inherently *bad.* What it is though is blunt: it's using a heavy, industrial, mechanical and quite expensive technology, involving pumps, turbines and water reservoirs that drown out parts of the countryside, to provide a fairly rudimentary valve for electricity demand, when we have at our disposal a light, digital, electronic technology that can respond in real time without human input in the most differentiated and subtle manner imaginable. It's a bit like knocking out a patient with a heavy object and cutting them open to remove a gall stone, when you could employ keyhole surgery...

VIII   SMART METERS

For a little while now, you may have noticed, smart meters have been a talking point. But there is not as yet any real consensus as to what exactly qualifies as a 'smart meter', what a smart meter needs to be able to do in order for it to be considered 'smart', and what the actual headline benefits of 'smart meters' are to the consumer. What most people agree on is that they are 'a good thing', that they have a variety of functions that allow you to know, for example, how much energy your household is using and what for, and what also seems clear is that there is a big drive towards them. In the

UK, for example, it is stated government policy to aim for every household to have a 'smart meter' installed by 2020,[CIV] and a great number and variety of energy providers and technology companies are embracing the concept.

Even Google, with its PowerMeter, made a brief foray into the technology, and although this did not last for very long, it offered up a fascinating insight nevertheless, because it signalled yet another shift, which is that energy is being taken out of the hands of specialist energy engineers, and put it into the hands of anybody who can handle and develop software, and that in turn is one of the really interesting aspects of this technology: instead of a few thousand energy specialists trying to cope with the challenges facing us, there are now millions of people working on them. Which means that, in energy, for the first time, important issues such as security, copyright, system integrity and functionality are all dealt with on the basis of a *creative commons,* and not on the basis of proprietary expertise. It is not all that surprising, for example, that Hewlett-Packard, an information technology company that most consumers associate with printers and PCs, for several years hosted an annual high level Executive Energy Conference, at which chiefs of the energy industry came together to discuss the future.

What we see manifest again and again is information technology reaching into and being applied in the energy sector, and so while smart meters still have some way to go before they fully deserve their name, we can expect that they will play a meaningful bridging role on the road to full system capability.

IX   A TYPICAL INSTALLATION

What then, might a typical installation in an average house or flat look like?

The schematic below shows the various elements that combine to turn a standard household into a 'smart' or 'intelligent' one, which is therefore system capable and can be connected to an energy network in the ways we've described: [CV]

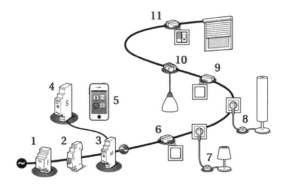

1, 2 – Filter and Main Switch

At the point of connection to the power grid or network, a mains switch allows for the electricity supply to be completely cut off. (This is already standard in any conventional set-up). A filter is used to detect and even out current fluctuations or distortions, and it also acts as an effective signal block to prevent electrosmog spilling over to neighbouring installations, or electronic eavesdropping.

3 – Meter

A smart meter replaces the ordinary meter of a conventional installation.

4 – Server

The server communicates with the network and runs the operating system as well as any applications.

5 – Smartphone Apps

Non-essential for the system to work, but extremely handy, smartphone apps allow the user to control and operate devices and appliances in their home from anywhere they happen to be.

6, 9, 11 – Built-in switch attachment

Ordinary switches for lights or blinds are turned into 'intelligent' switches with a recessed, hidden attachment.

7, 8 – Switch attachments

Switches for lamps and indeed any other kind of device can be connected retrospectively, by plugging the device into the attachment.

10 – Built-in device attachment

Fixed components that do not plug into a socket but are directly connected to the wiring can have their chip built in.

Obviously, the price of an installation like this will depend on a vast array of factors, such as the number of devices, how many of them will be connected to the system through external plug-in attachments, how many are being connected through built-in attachments, how much of it all is being done by a professional installer and how much just by the user themselves.

The important thing to note is that in order for the set-up to work, only really three components have to be either professionally installed or require a significant level of expertise: filter, main switch and meter. Much as in any ordinary household or office, the consumer is probably best advised to leave these to a qualified electrician. Everything else can be added and removed at will by the user without professional input.

THREE CASE STUDIES

VII

We've stated earlier in this book that we are not particularly keen on drawing up scenarios for the future. This has its good reasons: we live in a fast-moving world with an infinite number of unpredictable factors. So any scenario you paint is almost bound to turn out to have been wrong, as many a science fiction writer will be able to attest to.

But we want to demonstrate how, in practical terms, the concepts as well as the technology we are talking about might be implemented and how such an implementation could express itself directly and very noticeably on widely differing markets with genuinely life-changing benefits to entire societies.

So what follows are not predictions. The three case studies we present here illustrate possible and achievable outcomes. They are all based on what is currently doable with technology that's available now, and they are costed at 2012 levels, which means these are conservative estimates: photovoltaics are getting cheaper and more efficient all the time.

I   SWITZERLAND

Switzerland is a small, wealthy, landlocked country at the heart of Europe, known and respected for quality in architecture, design, high-tech engineering and gastronomy. Mountainous, multi-lingual and fiercely independent, it has a population of 8 million people,[CVI] over two thirds of whom live in towns and cities, which makes this a highly urbanised society that is also well educated, well catered for with social and cultural amenities, and well connected by transport and information links to the surrounding European nations and the world as a whole.

Primary energy consumption in Switzerland in 2015 amounted to 838,360 terajoules (TJ), which is the same as 232,878 gigawatt hours. Of this, just over half, 424,420 TJ, were oil products, and almost exactly a quarter, 209,690 TJ, were used as electricity. 142,150 TJ (60%) of Switzerland's electricity came from its many hydropower stations and 79,542 TJ (33.5%) from nuclear power. A little more than a third of all primary energy use (290,930 TJ) went on transportation fuels, of which 219,720 TJ (75%) were diesel and petrol, and 70,810 TJ (24%) were aviation fuels. Aviation thus made up about 8.5% of the overall energy consumption, pretty much in line with other European countries. (This is worth noting, not least because many an argument is brought against 'renewables' or 'sustainables', as they don't immediately appear to address aviation. But fuels used for air travel don't actually form that significant a slice of the energy pie, so giving them a lower priority for the time-being is indeed justified.)[CVII]

If Switzerland wanted to use photovoltaics to make itself effectively energy independent and continue its already chosen path of abandoning nuclear power, as well as gradually phase out its hydropower installations, it would have to cover 5.6% of its land surface with photovoltaic solar foils and panels:

Photovoltaic solar: surface requirement Switzerland to cover *all* primary energy consumption	
Photovoltaic surface required to generate 1kWh per day	3.5 sq m
Photovoltaic surface required to generate 1 MJ per year	1/360 sq m
Primary energy usage per year	838,360 TJ
Photovoltaic surface required to cover primary energy consumption	2,329 sq km
Land surface Switzerland	41,285 sq km
Percentage of land surface required to cover primary energy consumption from photovoltaic alone	5.6 %

These figures assume an efficiency of photovoltaic solar cells of 14 %. If their efficiency were to rise to 20 %, then the land surface required for Switzerland to cover *all* its primary energy requirement from photovoltaic solar would drop to just over 3 %.

It is not necessary for Switzerland to try and cover *all* its primary energy use from photovoltaic solar. Nor is it practical. The current infrastructure isn't geared to it, and with its Alpine terrain being so ideal for hydropower, and many of its dams and their lakes having become much loved features of the countryside, it would hardly make sense.

SOLAR ELECTRICITY TO REPLACE NUCLEAR But say Switzerland wanted to cover a third of its electricity use from photovoltaic solar. This, as we've seen, is the amount it currently generates in its nuclear power stations, and, like Germany, Switzerland has already committed itself to abandoning its nuclear programme.

Well, it would then have to find just over half a percent of its land surface for the installation of solar panels:

Photovoltaic solar: surface requirement Switzerland to cover one third of electricity consumption	
Photovoltaic surface required to generate 1kWh per day	3.5 sq m
Photovoltaic surface required to generate 1 MJ per year	1/360 sq m
Electricity generated by nuclear power	79,542 TJ
Photovoltaic surface required to cover electricity generated by nuclear power	221 sq km
Land surface Switzerland	41,285 sq km
Percentage of land surface required to cover electricity generated from nuclear power by photovoltaic solar	0.54 %

Switzerland is densely populated and there would not, at first glance, appear to be that much land available for solar panels. But leaving aside for the sake of this exercise city rooftops, motorways and rocky mountain slopes, the 0.54 % above does in fact represent still only about 1.6 % of agricultural land in Switzerland. Now, Swiss farmers receive around CHF 3.5 billion (USD 3.5 bn) annually in subsidies. If, starting in 2016, this money were allocated to photovoltaic solar, then by 2022, after 6 years and a total investment of USD 21 bn, the 60,000 Swiss farms could be providing all the electricity Switzerland needs to complement its hydropower generation, while at the same time earning themselves between $3 bn and $5 bn annually, which is easily the equivalent of their subsidy and about 40 % on top of their current annual income.

It is worth pointing out that using 1.6 % of their land for photovoltaic solar would not automatically reduce the farms' agricultural output by 1.6 % – even though they could easily sustain that with the extra income they'd be getting from the energy they'd now be selling into the grid.

But PV panels and foils are, as we have seen, very versatile and adaptable. They can lie on barn roofs, they can stand or hang suspended above the meadows with cows still grazing below, they can be translucent enough to cover greenhouses.

Switzerland is not the sunniest place in Europe, let alone on earth, by a long stretch, and food prices paid to farmers here are among the highest in the world. Even so, Swiss farmers could cover 1/3 of the country's electricity demand and at the same time increase their income by some 30 % if they were to cover just 1.6 % of their agricultural land with photovoltaics.

In this illustration, white fields give an approximate proportional indication of the amount of surface area this would take.

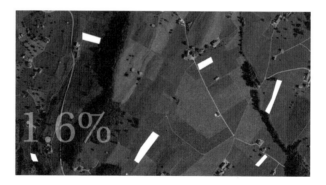

SOLAR FUEL TO REPLACE ALL FOSSILS USED FOR TRANSPORTATION

Swiss farmers can play an even bigger and more important part in their country's energy configuration. We have mentioned solar fuels on a number of occasions and we will be explaining what they are in just a little more detail later in this book, but they become highly relevant in two of our case studies, and so we need to touch on them in summary here.

There are various different ways in which solar power can be used to create artificial or synthetic fuels that are carbon-neutral and wholly compatible with the existing fossil fuel infrastructure.

A GENIUS PLANET

One of the most interesting and also promising technologies is being developed by German car manufacturer Audi, who call their version of synthetic methane 'e-gas'.[CVIII] In their case, it would be wrong and misleading to call it a 'solar fuel', because they in actual fact use surplus energy from wind turbines to take over-supply off the grid. The principle by which the fuel is generated, though, is exactly the same, and so whether wind or photovoltaic solar energy is used to make the fuel, the product is the same.

The process developed by Audi, and the one we cite here not because it's the only possible one, but because it's one that is being successfully trialled as we write this, uses electricity to separate water (H_2O) into oxygen (O) and hydrogen (H_2). It then combines the hydrogen (H_2) with carbon dioxide (CO_2) from bio-waste (farm manure, for example), which would otherwise pollute the atmosphere, and converts these two together into methane (CH_4). When this burns off, the byproducts are water and the same amount of carbon dioxide that has previously been bound into the fuel, which is why the cycle is carbon neutral. The resulting product – and therein lies its remarkable convenience – can be used with any normal car and for a multitude of other purposes.

What you need, therefore, in order to obtain a carbon-neutral, readily usable alternative to petrol from solar power is water, manure and some land to put up photovoltaic panels to generate clean electricity; in one word: farms.

With technology as it stands at the moment, the amount of land surface you need in Switzerland to produce the equivalent of 1 litre of ordinary petrol per year is 0.21 square metre:

Photovoltaic solar: surface requirement in Switzerland to produce 1l petrol equivalent	
Energy value for 1l petrol	42 MJ
Conversion efficiency of hydrogen electrolysis (turning water into hydrogen)	80 %
Conversion efficiency of methane synthesis (turning hydrogen and carbon dioxide into methane)	70 %
Energy required to produce 1l petrol equivalent (42 MJ / 1 × 80 % × 70 %)	75 MJ
Solar power yield per square metre per year	360 MJ
Land surface required to produce 1l petrol equivalent	0.21 sq m

This means that at today's levels, Switzerland would have to dedicate 3.6 % of its land surface to photovoltaic in order to cover its entire fuel requirement for transportation from solar:

Photovoltaic solar: surface requirement in Switzerland to cover entire fuel demand for transport	
Land surface requirement for 1l petrol equivalent per year	0.21 sq m
Fuel requirement for transportation per year	7,000,000,000 l
Land surface required to cover full fuel demand (turning hydrogen and carbon dioxide into methane)	1,470,000,000 sq m
Equals	1,470 sq km
Land surface Switzerland	41,285 sq km
Percentage of land surface required to cover entire transport fuel requirement from solar fuels	3.6 %

Again, looking at how much land Switzerland uses for agriculture (about 15,000 square kilometres), Swiss farmers would therefore have to dedicate just under ten percent (some 9.6 %) of their land to photovoltaics in order to wean their country off fuel imports for transportation. Assuming a total annual investment at the current farm subsidy level to of CHF 3.5 billion, but put towards solar energy 'farming' by exactly the same people, we estimate that by about 2020 to 2025 the price for farm-generated carbon-neutral solar fuel would be lower than the price of imported fossil fuels. From the outset though, participating farmers could approximately *double* their annual income, thus making pure subsidies superfluous; saving the taxpayer money, generating new revenues taxed in the country and therefore consolidating independence and growing wealth.

And indeed, the investment does not have to come from government or big global energy corporations: farmers, farming collectives, local councils, regional governments can all decide any time they feel like it that this would be a good idea, and go ahead with it.

II   INDIA

There were in India, at the end of 2012, some 600 million people without electricity; by 2015, according to the CIA World Factbook, this figure had dropped to some 237 million.[CIX] (Other sources vary this figure upwards a bit, but it seems safe to say that it is now below the 300 million mark.) The latest available comparison figures from the World Bank, which we've cited elsewhere in this book, tell us that the annual electricity consumption for the whole nation in 2013 was on average 765 kWh per head,[CX] though by 2016 this was reported to have pipped the 1,000 kWh mark.[CXI] There were vast differences between the highest consuming states (Dadra and Nagar Haveli with nearly 14,000 kWh per capita) and the lowest (Bihar with 134 kWh per capita).[CXII] 69 % of India's 865 billion kWh annual electricity use comes from fossil fuels, 2 % from nuclear, 17 % from hydroelectric and 12 % from other renewable sources.[CXIII]

By and large, it is cities where people have access to electricity and communication, whereas in remote rural areas the cost of extending supplies has so far often proved prohibitive. And indeed, the continuing drive to bring in the remaining 240 million people or so who aren't currently getting any electricity will be a massive undertaking: we estimate that the investment required by traditional means would be well in excess of USD 16 billion with a price per kilowatt hour of 8 cents. This does not take into account any environmental costs, which may be extravagant. A typical European coal-fired power station serves around 800,000 people. So we are talking about an additional 28-32 large conventional coal power stations, for example, plus the attendant infrastructure to get the coal to the power plant, and the electricity from the power plant to the end users, who are dispersed over vast areas of the sub-continent.

AFFORDABLE SOLAR ELECTRICITY FOR 100 MILLION HOUSEHOLDS

Using the approach and technology portrayed in this book, we believe it is possible, by contrast, to reach 100 million households (at current levels of rural occupancy in India, that is approximately 400-500 million people,[CXIV] so quite a few more than currently are without electricity) at a capital investment of just under USD 1 billion, and a mid-term price to the consumer of 2 US cents per kilowatt hour, with no adverse environmental impact, but wide-spread community stakeholding.

The key to this is 'smart' use of what's available and going about it in a modular way. Say you wanted to start with the most essential and useful thing for a farmer and his family in rural India. That would most likely be a refrigerator, light, radio and TV, a mobile phone, a computer and maybe a water pump. So leaving out, just for the time-being, high-wattage activities, such as cooking, hair-drying, heating or operating machinery, a power capacity of 360 watt would suffice to cover the basics.

One square metre of photovoltaic foil or panel in India can expect to generate around 90 watts. That's not very much and not enough to run a fridge. But if you had four square metres instead, you could already get going, because a fridge only actually operates around one sixth of the time. So if you had four households hooking up together in a village neighbourhood, each installing one square metre of PV, then you could run four fridges, plus some optional additions as above.
What, then, is in the 'starter kit'? Not a lot:
- 1 square metre photovoltaic solar panel
- 1 inverter
- 1 smart meter

- 6-12 identifiers such as the microchip explained above – they give each device an identity and allow them to communicate with the server to make sure, for example, that they don't overload the installation
- 1 server which orchestrates the electricity use for maximum efficiency

And, as you build from one household to a community of households, such as a whole small village:

- 1 distributor that hooks up individual households into a microgrid

An optional battery can be added to the kit. The reason this is not seen here as essential is that low voltage devices such as laptops and mobiles have their own batteries and can therefore be charged up during the day, while a small, modern fridge can retain enough of a low temperature throughout the night to keep its contents safe.

Total cost, then, to start with: USD 270, including the fridge.

The problem, of course, is that your average Indian rural farmer does not have 270 US dollars to spend. But what if this were approached as a community business proposition? What if a co-op were created, perhaps using the microbanking principle with investors from all over the world, or maybe government funded, or indeed any combination of investment sources with the money to lend Indian farmers the \$270 each they need to buy their starter kit and form small solar collectives of at least four at a time? The farmers thus become members of the co-op and pay back their microloan in the form of a low electricity tariff. Before long, these communities will be able to harvest not just the energy they themselves need, but in fact garner a surplus which they can sell first to each other or to other villagers nearby and ultimately into the grid, thus becoming stakeholders in their own and their country's energy supply network; and, with their loan repaid, they begin to share in the scheme's profit, allowing them to expand their energy generating operation.

How could this pan out in figures? Well, this is all pure hypothesis, of course, but let's run a case study of what might in fact be possible. And in doing so, let's set ourselves some ambitious, bold targets. This is neither vain nor insane: going about it the conventional route, say by planning and constructing nuclear power stations, you would have to be bold and ambitious too: a standard nuclear power station, even under next to ideal circumstances, takes at least 5 years and roughly USD 5 billion before it's ready to generate around 1,600 MW of electricity. So let's say: we don't have \$5 billion to invest, let's find a fraction of that to get going, in fact, less than a tenth, let's find some \$270 million.

\$270 million will equip 1 million households with a basic starter kit, giving about 4 to 5 million individuals instant access to basic electricity at around the normal market rate of 7-8 US cents, and generating an overall capacity of 89 MW, which corresponds to about 6% of the nuclear power station cited above. But that's in the first year, when the nuclear power station cited above is still in planning.

Now, we can safely assume that as soon as the first households in a village have electricity, others will want it too. So in the second year, we add to the nascent network, by continuing to sell bundled kits so that the network grows. In other words: people obtain a loan, use the loan to buy their own starter kit and hook it up. Because of additional demand, the unit price per kit drops to $240 and in some places there may now be small clusters of perhaps 20-30 units connected to each other, which would give them a power capacity of up to 2000 W, making it possible to start using more power-hungry devices and small machinery at times. This already markedly improves their economic potential and agility, two years into the scheme.

Let's assume then, for the sake of argument, that in the second year another 1.5 million units are installed, that's a 50% increase on the first year's rate of installation. This level of growth *is* ambitious, but it's not unusual in information technology. On average, with new solar panels now being added to existing ones, the output per installation rises by 25%, from 89 W to 110 W.

Our total output now is 245 MW which is about 11% of the capacity of a standard nuclear power station (bearing in mind that the standard nuclear power station we keep referring to by comparison still hasn't actually been built). But while the price of electricity from conventional generating methods will gradually go up or at best stay steady, photovoltaics, we know from every statistic there is, become cheaper over time. So in year 2, we can drop the price per kilowatt hour to 6.8 cents.

Year 3: we continue to expect rapid expansion, because the benefits of having electricity in your village really are compelling. Let's assume another 2.3 million people buy the starter kit. With additional solar panels installed, we now have 506 MW power, nearly a third of one nuclear power station (which by now, if everything has been going extremely well, is under construction). Whereas a conventional supplier would still have to charge 7-8 cents per kilowatt hour, our prices can drop to 5.6 cents per kWh.

The process continues, and as more people buy solar panels with a free fridge thrown in, and connect themselves to their growing local networks, the networks' capacity increases, as does their stability and robustness. Villages can start to connect to other villages. Soon, it's no longer a case of four enterprising farmers each with a solar panel, it's a case of entire farming communities augmenting their income through solar power; it's a case of workshops, schools, small businesses, local hospitals and nurseries being able to run everything they need, computers, mobile phones, internet access, refrigeration, medical apparatus.

Now connected and supplied with steady, cheap power, there is no longer the imperative to leave the country to get ahead: people can build a future where they are and put money back into their own communities. Let's therefore continue to be ambitious and assume a 50% growth in sales each year. By year ten, there will be 38 million new installations, bringing the total to over 100 million households with 400-500 million people: nearly twice the number currently estimated to be without electricity, in other words taking strain off other sources of power generation. The investment per household has dropped to just $30, while the overall investment amounts to a remarkable $930m. But that's still only one fifth of the cost of a nuclear power station, whereas the amount of electricity produced by farmers for themselves, for their villages and for sale into the grid as an actual revenue stream and wealth creator for their communities is equivalent to 3 nuclear power stations. Yet while conventional power would still cost around 7 cents per kilowatt hour, they are able to offer it to themselves and the rest of their country at 1.4 cents.

We are not going to pretend that this would all be simple, smooth and straightforward. But there are some extremely important points to this case study:

1) Solar is often dismissed as expensive, *too* expensive at any rate for developing countries, when in fact compared to conventional methods of energy generation with their attendant high infrastructure costs, photovoltaics is remarkably cheap, and getting cheaper

2) The process for a scheme like this can commence immediately and with almost instant benefit to the end users, rather than being reliant on massive infrastructural and planning-heavy projects: it is ready to go

3) Building end users into the scheme as active participants and stakeholders makes it economically viable and desirable, because it creates wealth directly at grassroots level

Since we first drafted this hypothetical case study, programmes to implement schemes very similar to this have sprung up in India as elsewhere, and as you would expect, they show some successes, and also of course highlight some challenges.[CXV] What is most promising though is that the approach in principle is sound and works, and major players, such as Tata Solar Power, are actively pursuing it.[CXVI]

III   ETHIOPIA

Landlocked like Switzerland, but situated on the Horn of Africa, with Somalia and Eritrea as neighbours, Ethiopia provides a stark contrast to the Alpine State. With a size nearly twice that of Texas or just over twice that of Spain, it has an estimated population of 102 million, of whom more than 80 % live in rural areas, less than half (49 %) are literate and more than three quarters – some 71 million people – have no electricity. These statistics take their toll: with 62 years, Ethiopia's average life expectancy ranks 194th in the world, compared to Switzerland's 83 years, which rank it 9th.[CXVII]

Not having electricity, for Ethiopians, is not an inconvenience, it's an ongoing calamity. But this is exacerbated, in large parts of the country, by underdeveloped infrastructure and a lack of clean alternatives to traditional fuel. About 1,400 square kilometres of land – nearly three times the area covered by Ethiopia's capital Addis Ababa – is deforested each year, as the vast majority of Ethiopian households continue to use firewood for cooking.[CXVIII] The devastating effects this has on the environment – land erosion, desertification – and the attendant carbon emissions are well documented and understood. In addition though, these domestic wood fires expose millions of mostly women who stoke them in the preparation of meals for their families to indoor smoke every day of their comparatively short lives. So women in rural areas have an incidence of respiratory diseases three times higher than their sisters in urban areas.[CXIX]

Finding viable substitutes for firewood is therefore one of the highest priorities for this country. And of course there are many other areas in which development would bring tangible benefits. Less than 14 % of the road network, for instance, is paved, which means that getting from one village or town to another or to the nearest town or city – all essential for trade and business – is slow, cumbersome and dangerous.[CXX] As of 2014, there were in Ethiopia 3 motor vehicles per 1000 people: in rural areas, the standard means of transport are pack and draft animals.[CXXI] Supporting Ethiopia's road building programme is therefore another high priority.

Both these issues, fuel for cooking and fuel for transportation, can be addressed with synthetic methane, exactly the same type of technology as we have seen in the case study for Switzerland above. And electrification can be achieved with a model very similar to the one we have postulated for India.

Let's play through some numbers for electrification first. Here, the benchmark against which any programme or initiative needs to be measured is the Gilgel Gibe III dam, which went into operation in October 2015 with a generating capacity of 1870 MW.[CXXII] It will form part of the Gibe 'cascade', a series of hydrodams on the river Omo in Southwest Ethiopia, which also consists of Gibe I, with a capacity of 184 MW and Gibe II (420 MW). Further expansion plans include Gibe IV (1472 MW) and Gibe V (560 MW).[CXXIII]

Gibe III cost an estimated USD 1.8 billion to build and effectively doubled Ethiopia's electricity generating capacity at a stroke. But with much of the country lacking any type of infrastructure, rural communities will not benefit from this development.

SOLAR ELECTRICITY FOR A SMALL TOWN   Take Leku, for example. It has a population of 15,100, and sits about 50 miles south of the Abidjatta-Shalla National Park, far from any major conurbation.

As we write this, it has no electricity, so if we were to take the – admittedly very modest – average Ethiopian electricity consumption of 65 kWh per capita as a starting point, we would need 4,260 square metres of photovoltaic solar cells to power the entire town. (4,260 square metres, just for illustration purposes, is about the size of a standard association football pitch.)

In 2015, Ethiopia's GDP per capita was USD 486, USD 1,530 when adjusted for Purchasing Power Parity (PPP). So it would be necessary to offer a solar starter kit similar to that used in our India example, but at considerably less than the $270 cited there. This can be achieved by taking account of lower distribution costs and a more favourable exchange rate, for example, and leaving out, to begin with, the fridge. Working, therefore, on the assumption of a starter kit cost of $115, the repayments for a household over five years, including interest on a microloan, would be about $37 per year.

A starter kit, much as described above, provides 94 W of electric power, which is enough to operate a radio and a water pump and charge a phone or a laptop. This may not sound like a lot, but it is immediate and, at just over three US dollars a month or around ten cents a day, affordable. And as we have seen before, the moment the first clusters exist, the amount of power that becomes available goes up and the price comes down.

After a year, the cost of the starter kit – through economy of scale – has gone down to $95 and annual loan repayments have come down to $30. As in the Indian example, small networks start to build: ten households can now hook up their installations together and jointly produce 940 W for their community, enough to start running fridges and some apparatus.

After another year, the cost of a starter kit comes down to $78 and the corresponding annual loan repayment is down to $25. A community of 30 households can have a maximum load of 2,700 W, and collect enough power to sell surplus energy to their local neighbours.

Year 3: The cost per starter kit is $64 and the cost of credit per year has come down to $21.

Again, as with the case study given for India, rural communities can now form actual power networks and generate an income for themselves, selling electricity on to nearby villages, ultimately into the national grid.

SOLAR FUEL FOR A VILLAGE   A slightly more ambitious, but nevertheless interesting proposition comes with solar fuel. Irrespective of whether Ethiopians in rural areas get electricity or not, and we clearly advocate that they do, what is urgently needed are clean alternatives to wood fires for the purpose of cooking, and a way to accelerate motorisation without bringing high financial or environmental costs into the equation.

So imagine a small Ethiopian village consisting of, say, 80 households. With one litre of solar fuel per day, each household can cook for two hours, travel 30 km by scooter or tuk-tuk (for example to the next village and back), or illuminate a room during the night for 5 hours. Again, it's not much, but it's a start, and it's better than spending hours each day scavenging for scarce fire wood and then burning that inside the hut.

In order to produce 80 litres of solar fuel a day to give each of the 80 households in our hypothetical village their one litre, you need, on the Horn of Africa, some 1,360 square metres of photovoltaic solar surface.

Plus you need a small solar gas synthesiser to actually make the fuel. With the participation of a well-positioned technology partner – Volkswagen, Toyota, Tata, for example – we estimate that at large enough numbers, say one million units, the cost for such a piece of kit could be brought down to around $1,390. This is, in effect, a target figure: the level at which a scheme such as the one suggested here becomes viable.

Because now, over a ten year period, a microloan for their equipment – solar panels and solar fuel synthesiser – including interest, would come to about $310 per year. That seems prohibitive, when your annual GDP per head is around $486. But say you get just ten out of the 80 households working together and pooling their resources, they will be able to, at a cost of $31 per year each, produce fuel for themselves and for the other 70 households in their village, selling the fuel to them at a low cost and quickly returning a profit on their own investment, allowing them to expand and step up production. Once again, you achieve an immediate effect at local level, involving the people you are helping not as recipients of aid but as stakeholders in their own development, as small scale, environmentally sound, entrepreneurs: you create actual wealth and the benefits this yields.

Giving rural communities in a country like Ethiopia access to electricity, to the internet and to transport transforms their lives very quickly. Electricity can power water pumps, so you can better and more efficiently cultivate the land: you have food. Once you have a mobile signal you can go online, once you can go online you have access to education and e-commerce.

The reason big cities or city-like agglomerations in developing countries struggle so much with overpopulation and slum conditions is that they are magnets of – often still very limited – opportunity to young people who seek wealth, education and fulfilment, not just for themselves but for their own prospective families. And indeed health care: very often the only place where you can obtain medicine for a sick relative is in the city, and if transport is slow and difficult then perhaps the only place where you can live with your ailing grandmother is in or near the city.

But if economic opportunity, education, goods and services become available where you are – because you have electricity, you have the internet and you have cheap fast transport when you need it – you no longer have to move to the city, you can, as we've seen, lead a contemporary urban lifestyle (or at the very least begin to approximate one) right where you are, in your local village. This may not be enough for everyone, but for many it will be. And so not only do you markedly improve the quality of life and the life expectancy of your rural communities, you also ease pressure off the cities, allowing them to grow and adapt at a more organic, altogether more wholesome and, indeed, sustainable pace.

THE BIGGER PICTURE

VIII

The energy issue, as is abundantly apparent, does not stand in isolation, nor is it about simple or simplistic solutions. It is not about 'alternative' energies, and it's not about 'going back to basics', let alone 'back to nature'. It's about accessing the principal source of energy we have, the sun, and a multitude of secondary sources that are mostly, if not all, solar energy in the form of weather. It therefore also isn't about 'renewables', because the sun and the weather, in common with geothermal and tidal power, are not 'renewable' sources, they're inexhaustible sources of energy. And it's not about using this energy in a localised, primitive way: it's not about solar cookers, fun though they are, or fridges that stop working when the sun goes down.

What the energy issue is about is energy logistics, energy networks and the convergence of energy and information technology with networked energy applications. And so it is about a seismic shift at every level: energy generation, energy distribution, energy pricing, energy trading, energy consumption, energy storage. Which is why the recent words of an unlikely figure, uttered in a very different context, ring especially true:

David Hockney,[CXXIV] the great British artist who was born in Bradford in 1937 and who spent many years living in California before returning permanently to his native Yorkshire in the 1990s, was invited by the Royal Academy of Arts in London to put together a big show for the first half of 2012.[CXXV] By now in his seventies, Hockney had started using the iPad to create paintings which would then be enlarged and printed out for display at the gallery. This, unsurprisingly, raised an eyebrow here and there in more traditional art circles, but it also gained him the admiration of many, because rarely had anybody seen a grand old master of oils and acrylics adopt a mainstream consumer gadget like this and turn it into a professional tool to such compelling effect. His explanation, given in interviews for the exhibition's audio guide, was that the iPad allowed him to quickly put together a colour palette and then work very fast, on location. "What technology needs," he says, "is imaginative, mad people to start using it." And that's exactly how we feel when it comes to energy.[CXXVI]

I   THE YOGHURT PHENOMENON

Let's venture a rare prediction here: fifty years from now (let 'now' be a hot and sticky humid day in early August 2011, when we first wrote this particular sentence), the way in which people all over the world handle and perceive energy will be incomparable to the way they did fifty years ago. Fifty years ago, in August 1961, John F. Kennedy was president of the United States, Britain applied to join the EEC and, on the 13th of the month, construction of the Berlin Wall began: the most potent symbol of the Cold War and the division between East and West. If a week is a long time in politics, then fifty years are an eternity, and the same is true for technology. But in *energy* terms, not much really has changed between about then and about now. But between about now and about another fifty years down the line, many of the significant changes we're only just seeing the beginning of are going to bear fruit.

Also about fifty years ago, between 1961 and 1967, the Boston and Maine Railroad in the United States was testing a system called KarTrak, developed by one David Collins,[CXXVII] to – much as the name suggests – keep track of its cars.[CXXVIII] Young Mr Collins (he had only recently graduated from the Massachusetts Institute of Technology, MIT) was setting a marker not only for the deliberate misspelling of technology names, but also for the automation and digitisation, you could call it the virtualisation, of logistics. Because KarTrak was like a prototype for what we soon began to recognise universally as the barcode.

The Universal Product Code (not the only, but probably still the most familiar barcode there is) started being rolled out in the early seventies, and shopping for yoghurts has never been quite the same again. Nor has it been the same for anything else. But yoghurts provide us with a particularly satisfying example for the point we're about to make, because yoghurts embody much of what is true about many aspects of our consumer experience in the past and now.

First of all, yoghurts are nothing new. The earliest mention of yoghurt dates back about 2500 years, to Indo-Iranian records, where, in combination with honey, it is referred to as 'food of the gods'. If you have ever been to Greece and had the typical Greek yoghurt with honey there, you will know that this description is exactly accurate. It is believed though that yoghurt or yoghurt-like dairy produce was used as early as 2000 BCE, maybe earlier.[CXXIX]

Secondly, much as we love yoghurt, and we patently do, and much as we have reason to believe that yoghurt is good for us – something that was first expressed by the Persians who attributed Abraham's longevity and fecundity to his regular intake of the stuff, and there is plenty of evidence to support health benefit claims for yoghurt – we don't actually need it. We can get by very well on a diet completely devoid of yoghurt, even if many of us would, at least in a melodramatic fashion, rather be dead than having to do so.

Thirdly, although it keeps reasonably well when chilled, being a dairy produce it is best and most safely handled and eaten fresh.

All of these factors make it an ideal candidate for us to illustrate how the bar code brought us plenty through logistics. What is the barcode? It's a way of identifying each individual product as a unique class item. The bar code 5 201054 016442 stands for a small (170g) tub of Fage Total natural yoghurt, this one with a best before stamp of 09.03.12, and a very small chance indeed of lasting that long as this sentence takes shape...

The simple act of identifying individual products revolutionised everything associated with the production, transportation, distribution, sale and consumption of everything, not just yoghurts, because now it was possible to easily, automatically and continuously keep track of what comes into and what goes out of a system, and thus keep the system going at a very different level.

Whatever you make of Walmart or Aldi or any other supermarket chain, the reason they can exist and the reason they work is the barcode. With the barcode you can have just-in-time delivery, which means you don't clog up warehouses. This in turn means you can offer a multitude of choice on perishable goods because the goods will not be stacked somewhere and rot, they will go from the production line via short term cold stores to the shelves and into the consumers' fridges, easily within 48 hours. What barcodes did to yoghurts, and with yoghurts to everything else, is that they made production secondary to distribution, because suddenly the question was no longer, is there a factory, a farm or any other supplier nearby who can provide this product, the question was simply, how do we get this product from wherever it happens to be produced to the consumer.

And so it's thanks to this, automated, real time, digital logistics that we have not two or three types of yoghurt to choose from today, but *dozens.* This is possible only because now you know exactly how much is being bought when, and you can make sure that even if only one in a thousand customers just *love* their ultra-low-fat lychee in drinkable form yoghurt, you can have it ready for them, and because you can have it ready for them they will come to you and buy not just their favourite yoghurt drink but also all their other groceries, which may amount to two hundred dollars a trip.

And in what way is this relevant to our energy situation? You may well ask. The barcode story serves as a pertinent example of what happens when you sort out the logistics. Sorting out the logistics for consumer goods meant that supply, quantity and *production* were no longer the issue. And this is precisely where we are at in terms of energy now.

There is another dimension to this, though, which is no less important. And that is one of choice and quality. Excepting, for the sake of this argument, specialist trade and highly sophisticated suppliers – such as, for example, a local delicatessen, or Harrod's Food Hall – the main concern of grocery stores far and wide had been to meet basic demand. As far as the *mass market* was concerned, the options for what you could put on your shelves were comparatively limited, mainly by seasonal and other market factors, such as what produce is available what time of year and at what price. You could argue, therefore, that your standard food store was not entirely dissimilar in some respects to your standard mobile phone handset: it provided you with a range of products/services, and as long as they met your basic demand, you and your supplier were reasonably happy. Much as it did not occur to you to ask your local Coop for a Greek style yoghurt that you could drip a spoonful of honey in for your breakfast (and thus feel like a god), you wouldn't ask your mobile phone service provider to give you access to traffic updates.

It's not that either of these were inconceivable or impossible, it's mainly the case that the amount of effort and cost someone would have to go to to make these things available to you would have been in most cases prohibitive. Which is why a specialist delicatessen or a fine goods supplier to the very well-off would maybe not have batted an eyelid, but your ordinary corner shop would have had to simply say 'no'.

But by sorting the logistics, you make available not just what's there in your immediate catchment area, but what's there anywhere in the world. In the case of yoghurt, that's due to the barcode. Because obviously it was always possible to shift goods from one end of the globe to another. But what makes doing so economical, quick, reliable and scaleable is the representation of your item as a digital code. It is, in a sense, making it virtual. And so channelling the yoghurt from the manufacturing plant to your check-out till happens in the physical world with the actual tub of yoghurt, and it happens at the same time in the virtual world with a string of numbers representing that tub of yoghurt.

It is no coincidence that what turned the internet into the staggering success that it is, is not the wealth of information available to us in itself, although this has clearly greatly increased as a result of it being such a success. What turned the internet from something that as late as the late 1990s to many looked like it was 'just for geeks', into something that hardly anybody can do without, was that somebody — first Yahoo and then with considerably greater impact Google — sorted the logistics: how to get at the information and channel it from where it is to where it is needed. With information we call it indexing. Barcodes are nothing other than a way of indexing products.

So sorting the logistics gives you access to plenty. And once you have plenty you have choices. This may strike us as a mere luxury, but in fact it is more than that. It is a defining aspect of civilisation.

You do not really speak of a food culture as long as everybody is concerned only with having enough on their plate to survive. The moment people can vary their diet, add spices here and there and experiment with ingredients; improvise with combinations and invent new methods of preparation — in other words: broaden the spectrum of their options — you can have a food culture.

As we are writing this, one of the most catastrophic famines is taking hold in Western Africa. Naturally, aid organisations are doing their utmost — and in difficult circumstances — to get food and water to the stricken areas, a task that is massively hampered by civil war in Southern Sudan. It is fair to say that the absolute top priority, really in any meaningful way the *only* priority, is getting supplies to the people there, so they won't die. 'Choice' does not come into it. Of interest is a baseline cleanness of the drinking water and edibility of the food that's being taken there. This is the extreme end of the scale, the emergency supply end. As far as we can currently imagine, we in the wealthy parts of the world are pretty much at the other extreme end, where a trip to the average supermarket involves a plethora of choices, which are in the majority completely unnecessary. It is not *essential* for our survival, either as individuals or as a species, to be able to choose from 300 different types of yoghurt. That's the exact number of yoghurts (not including drinks) available at the time of writing, in 9 separate categories, on the Tesco website. Even allowing for the possibility of some, perhaps even many, of these yoghurts featuring in several categories at once – who's to say, for example, that a 'Kid's Yoghurt' should not also feature in the 'Daily Health Yoghurts' category, and we haven't carried out a check – you can still say with absolute certainty that we are living in yoghurt heaven. We have no need for more. We are saturated with yoghurt, and spoilt for choice. And that's not a bad position to be in, because it means the chances of you being able to eat *exactly* the kind of yoghurt you like for breakfast, even if it is a radically different one to the kind of yoghurt you like to eat before going to bed, are not just fair, they are amazingly good.

II   THE APPLICATION PRINCIPLE FORETOLD

But can this really be compared to energy? And quite apart from that: couldn't we be just as happy and content if we only had one or two yoghurts to choose from? Aren't we *overburdened* with choice? And how exactly does it benefit people in the Sudan if we can choose from 300 different yoghurts; wouldn't a fairer world have us being able to choose from maybe 15, and them being able to choose from as many?

The answers to all these questions go, surprisingly perhaps, hand in hand. Because the true significance of all this, of course, is not the specifics surrounding our choice of yoghurts. Nor those relating to any other consumer product. Who ever knew that there is a difference marked enough between a 'Daily Health Yoghurt' and a 'Diet & Low Fat Yoghurt' for it to warrant 38 types in one category and a remarkable 65 in the other? How could this ever be of any importance to us at all? And herein lies the crucial point. It *isn't!* That's the beauty of it. Because in this, as in many another context, our concern has long ceased to be what's *necessary.* Our concern is largely now what's *possible.* And in that sense you can draw an interesting and valid parallel to what we think the energy discussion is going to be about, which is *applications.* Applications are not a function of what's *necessary* either, they are generators of possibilities. And beyond possibilities, opportunities. For everyone, *irrespective* of where in the world they are, whether they're rich or poor, whether they have been brought up to feel privileged or not.

So you could almost go as far as to say that the situation we have brought about with yoghurts is an urban application of the availability of milk. Once upon a time, in a rural agricultural society, milk was a staple source of nutrients and fat, and the motivation for turning it into anything at all was primarily to keep it edible. With modern farming methods and, far more to the point, contemporary logistics, milk has become the baseline ingredient for a myriad of yoghurts, cheeses, butters and drinks which help us, within our busy urban lifestyles, to keep healthy and fit.

This same principle applies to the way we use almost any other type of food, clothing, even cosmetics or medicine: these are all to a greater or lesser extent essential for us, but we don't treat them as such, we treat them as sophisticated, complex and highly nuanced facets of our lives that, far beyond simply keeping us going, express who we are, reflect our culture and enhance our experience of living to the point of sheer luxury. Luxuries, indeed, which would once have astonished kings and emperors, but which today someone on unemployment benefit may take for granted. And herein lies the underlying value to society as a whole: logistics, choice, an abundance of yoghurts, none of these are for the lucky few. None of these are intended, designed, invented for an elite. True, we, compared to the people of Sudan right at this moment, *are* the elite. But not by definition and not by necessity either. Everything that is possible for us today is also possible for everybody else on the planet, if not today then in an attainable future. Not in spite of logistics, and technology, but *because of it.* We can look at where we are today and say, truthfully and realistically, we have a long long way to go yet before there is wealth, health and freedom for everyone on this planet. We can also look at where we are today and say, as truthfully and as realistically: we have come a long long way already.

Elsewhere in this book we compared levels of literacy, life expectancy and energy use at different times in history and in different parts of the world. We could draw them all together now and present you with several pages of impressive figures, tables, pie charts and delightful graphs, but that would rather derail us from our track and take us on a detour that is perhaps not strictly necessary. You can look up any set of relevant metrics relating to quality of life and compare today to a hundred, two hundred, five hundred and two thousand years ago and you *will* find, over and over again, that never before in human history have so many of us had it so good.[CXXX] We *know* that we mustn't rest before we *all* have not just enough, but plenty, as well as human rights, potential for self-fulfilment and opportunity. So we're not saying 'job done'. In fact, we're saying the opposite: 'There's big work ahead, and the only way to go about it is to avail ourselves of technology.'

The microchip in an electric device has a similar function to the bar-code on a product. It facilitates a new kind of logistics. And when we sort the logistics of energy, we will have a comparable situation as we have with yoghurts: we will no longer need to concern ourselves with the baseline supply. It will not be of interest to us that there is electricity to start with. What will be of interest to us will be what we can do with that electricity. We will, very simply, be able to *cultivate* our energy experience. And this we will do with applications; exactly as we are cultivating our internet or mobile handset experience with applications, so we will do with energy. And this is why we are saying that you can expect the uses you put energy to and your overall 'take on energy' to change as much between now and the middle of the 21st century as changed our 'take on food' over the last fifty years or so.

When you start mentioning people who are starving, you touch on something raw and serious. So let us repeat and emphasise: there is no need for this. Our contention, here, that we can have a world of plenty – be it yoghurts or kilowatts – and with that plenty a plethora of choice, is not restricted to the happy, lucky few at the expense of the many. It's not something we can only afford because we exploit the developing world. True, this does happen: the way much of our current economic systems are structured lead us, wittingly or unwittingly, to enjoy privileges and luxuries at the expense of people who are worse off than us. But this is neither a characteristic of plenty nor is it inherent in science and technology. Again, the opposite is the case: with science and technology, with networked logistics, with an applications rather than resources based approach, there is absolutely no need for either starvation or exploitation. The idea is to lift *everyone* out of an existence which scarcely covers the basics, and onto a plateau where we can *all* shape our lives creatively and play to our human strengths.

IV   PRINTED ENERGY: PHOTOVOLTAICS

Another thing that we need to talk about in some more detail now concerns not the logistics of energy but the printing of it. We're being a little provocative here, because the action is really one of garnering or collecting or catching energy. None of the verbs quite does it justice in this context, because putting up photovoltaic solar cells to face the sun and having them convert energy into electricity is still not quite the same as harvesting something that's been growing, or hunting something that's running around and catching it, or gathering things up that lie dispersed around the countryside. We may need to invent a new word for what it *really* is that solar foils do, but for the time being we're happy to leave that to one side and concentrate on how we make solar cells in the first place, because the actual *process,* when it comes to photovoltaic solar, lies here. Once the cells are up, they 'do' very little, though the little they do is of tremendous value, as we realise.

The reason photovoltaics are of such interest to us is because photovoltaic solar cells turn solar energy directly into electricity, without any intermediate steps, without any mechanical parts, without the use of any further materials or fuels, without any other conversion processes. The way they do so is relatively simple; the way they are made in order to be able to do so is relatively complex, which is why photovoltaic solar cells are extremely easy to install and virtually maintenance free, why they are cheap to produce in very large quantities, but expensive and time-consuming to develop.

Photovoltaic solar cells can be manufactured as sturdy panels or as very thin, flexible foils, they can be produced in any shape or size and strung together to cover vast areas and feed power into the grid, or they can sit facing the sky individually, attached to some electricity consuming device. They offer, therefore, maximum versatility and minimum environmental impact, once they're in use. And they are almost completely recyclable.

So how does a photovoltaic solar cell turn solar energy into electricity?

THE PHOTOVOLTAIC SOLAR CELL   The sun, as everyone knows, releases an enormous amount of energy at exceptionally high temperatures. At the core of the sun, atomic particles are melted at around 15 million degrees Celsius, and the energy that this ongoing gigantic nuclear reactor releases comes to us in the form of heat, light, ultraviolet rays and radiation. Luckily for us, relatively little comes as radiation that may be harmful, and a lot comes as heat or light, both patently useful to our existence, as long as we expose ourselves to them in reasonable doses.

Light has the extraordinary quality that it 'consists of' rays at the same time as atomic-sized particles: photons. (This is known as the 'wave-particle duality', and it's one of the many fascinating aspects of light, that we could go into great detail about, but won't.[CXXXI]) Much as heat and light rays are energy, so are photons, and photons make themselves felt when they reach our planet by interacting with the atoms that are already here. You can see and feel this in action if you park a black car directly in the sun: it will get hot. Given enough sunlight – say around midday on a summer's day – it can easily get hot enough to fry an egg on the bonnet. But why exactly does it do that?

As you will remember, almost all atoms consist of a nucleus that is surrounded by electrons. The nucleus, you may recall, consists of protons and neutrons, of which the protons are positively charged and the neutrons have no electric charge (the exception to all this, as we've already mentioned, is the hydrogen-1 atom, which has no neutrons). The electrons that surround the nucleus are negatively charged and you will also recall that they are incredibly small.

What happens when a photon hits a surface, such as the metal of your car, is that it displaces one or several electrons in the atoms that the surface is made up of. The extent to which it can do so depends on the properties of the surface, which is why some materials get much hotter than others. Factors that play an important part are the material's reflectiveness (a white car, you will notice, gets less hot than a black car, which is not because of the material the car is made of, but simply because white colour has a much higher reflectiveness than black and therefore sends many more of the photons back into the surroundings rather than allowing them to interact with its own atoms). The electrons thus displaced are no longer attached to a nucleus and therefore do what to them comes naturally: they whizz around until they find a new 'empty' spot on either the same or any other atom with similarly displaced electrons where they can once more attach themselves and settle. This movement is the energy carried by the photons converted into what we register as heat.

Instead of heat, the photons' energy can also be converted into electricity though. In order for this to happen, there need to be in place some fairly specific conditions, but how these conditions are achieved is relatively flexible: there are a whole range of materials, and several ways of putting them together, that can facilitate the same or very similar processes, but the principle is more or less always the same.

Currently, most standard photovoltaic solar cells are made of silicon. The reason for this is that silicon lends itself reasonably well for fashioning into an arrangement that allows for the conversion of solar energy directly into electricity, and its raw materials are available the world over in plentiful supply: sand or quartz. Quartz has the advantage over many other types of sand that it is already relatively pure: it consists mainly of silicon and oxygen. ('Ordinary' sand such as you might find on a beach also consists of silicon and oxygen, but it may also contain traces of any number of other minerals, for which reason the effort of getting at pure silicon from it is normally much more laborious.)

For our purposes in a photovoltaic solar cell, the useful component is silicon only, and so whether it is quartz or any other kind of sand that's used, the important thing is to get the oxygen and any impurities out of it. With quartz, as it has nothing but two oxygen atoms per silicon atom, this is comparatively straightforward: you grind the quartz up into a fine powder and melt it. What you get is liquid silicon which contains no more oxygen. When this cools down, it once more solidifies into hard rocky pieces, which now are 'raw silicon'. This then gets heated up once more and thereby liquified and in this liquid form relieved of any remaining impurities, and when it now settles into dry blocks again, it is pure silicon with nothing else in it.

Silicon, when pure, crystallises into an impeccably regular cubic structure that arranges itself in a precise grid pattern, whereby each nucleus is surrounded by four electrons. This is handy, because it makes for an ideal base for our photovoltaic solar panels, as we shall see. Before it can come into use though, this regular atomic structure gets mixed up again by deliberately adding a new substance to the silicon: boron. Boron is a metalloid semiconductor and its atomic structure is at variance from that of silicon: it has one nucleus, surrounded by only three electrons. As the boron melds with the silicon at very high temperatures and in liquid form, the material that was pure silicon now loses its perfect grid structure and by the time it dries into blocks again it now has changed: in some places where before there was a silicon atom, there now is a boron atom, which means that parts of the grid now have an electron missing; where before there were four, there are now only three.

At this stage the silicon blocks, now containing boron, are sliced into thin 'wavers'. These are washed and then sent through a process which burns a layer of phosphorus to one side on each waver. This is the bit that resembles the 'printing': if you were to consider your silicon-boron plate or waver the page, then the phosphorus is the approximate equivalent of the chemicals that turn paper from ordinary into photographic paper and allow a picture to appear on it. Here, however, it is not the appearance that matters – this only changes marginally – but, again, the atomic structure of the new layer: phosphorus atoms consist of a nucleus surrounded by *five* electrons. This means the plates or wavers now have two layers with distinct electric charges: the silicon-boron layer has fewer electrons (either four or three to each nucleus) than the phosphorus layer (which has five). As electrons are negatively charged, this means that the side with the phosphorus on it is negatively charged compared to the other side, which is positively charged.

The photovoltaic solar cell is practically ready. All that gets added to it now is an non-reflective coating to maximise the amount of sunlight that stays on each cell, a thin wire grid to allow for the transfer of electrons from one layer to the other, and connection points to daisy chain cells together and hook them up to an electric load, which can be anything from a tiny LED lamp to a big piece of machinery, or indeed the electricity supply grid. (If the solar cell is connected to the grid, it has to go via an inverter that transforms the charge from the solar installation's low voltage – usually about 12V – to the grid's high voltage, which is usually 110V or 230V. At the same time it also converts the charge from direct current to alternating current and brings it in synch with the grid's 50Hz or 60Hz frequency, again depending on where you are.)

When the solar cell now is exposed to sunlight, the photons once again dislodge electrons in the surface they encounter. But unlike the bonnet of a car, where the moving electrons have nowhere to go other than to another atom in the same layer and where the energy that their movement releases is converted into heat, here now they have an obvious route: the thin hairline wires burnt into the cell allow the electrons to travel from the layer where there are many of them (the negatively charged phosphorous side) to the one where there are few of them (the positively charged silicon-boron side). On the way, they can power the electric load that the wire is attached to: the circuit is complete and we have electricity.

One of the things that makes photovoltaic cells so interesting is their extraordinary versatility: everything we've described here is just the principle. The actual materials used can, and do, vary a great deal. The 2010 Millennium Technology Prize for example, awarded by TAF, the Technology Academy of Finland in Helsinki, went to the German-born, Swiss inventor Michael Grätzel [CXXXII] for his work with dye sensitised solar cells (DSCs). [CXXXIII] Instead of silicon, they use titanium dioxide which, in spite of its name, is a readily available, inexpensive pigment used in paint. That notwithstanding, DSCs can also be made completely transparent. This means it is possible to use entire glass surfaces, such as the facades of large office blocks, as energy garnering solar panels. While DSCs, at between 11 % and 15 %, have a low efficiency compared to silicon cells, they are cheap to produce, flexible and adaptable, and they work even in cloudy or indeed rainy conditions. They therefore open up another new set of opportunities in engineering and architecture. [CXXXIV] And it is now not only possible to make photovoltaic solar cells on thin flexible film, it is also possible to make them from organic materials, with increasingly useful results. [CXXXV] All of which means that the main caveat against photovoltaics, that they use natural resources in their manufacture and that the manufacturing process is in itself highly energy intensive, becomes steadily less significant. If it is possible to *grow* the building blocks of a solar cell, then there really is a new horizon opening up as to their potential.

You may remember us talking about 'always on' energy and using the example of a photovoltaic foil at one end and an LED foil at the other end of a cable. Once you have an energy network and infrastructure in place, this becomes a simple plug-and-play reality: I can plug a photovoltaic solar cell into my normal domestic power socket at home in London (again, via an inverter), and while the sun is shining on my roof, I can plug a fan or a computer or a TV or anything I feel like into another socket and it will run on the sun's energy. And if I don't want or need this energy, someone else, in my street, anywhere else in London, anywhere on the island, really, in fact anywhere in Europe can plug any electrical device into their power socket and run it with the sun's rays that stroke my roof. All that's needed, therefore, in principle, for everyone to have constant access to cheap, clean power, is enough solar cells across many and varied locations, and a network to connect them.

V   THE POWER OF EXPONENTIAL GROWTH

In 1985, the average capacity of a personal computer hard drive was approximately 0.01 gigabyte or 10 megabytes. Fifteen years later, by the year 2000, this had increased by about a thousand-fold to approximately 10 gigabytes. By 2015 it had increased by about a hundred times on that figure, to 1,000 gigabytes or one terabyte. So over the course of thirty years, your average hard drive capacity has increased not three or four times, not ten or a hundred times, but *a hundred thousand* times. It has, very simply, doubled every two years, exactly as Gordon E. Moore predicted in 1965. At the same time, the price we pay for a computer hard drive has tumbled.

When DVDs first came to market in the early 1990s, the unit price to the consumer was several dollars. As we write this, a box of 100 recordable discs can be had for about twenty dollars: twenty cents apiece. Today, we can store on one speck of semiconductor material no bigger than the full stop at the end of this sentence the amount of information that would once have taken up the great Library of Alexandria.

Exponential growth is extremely potent. The most famous story to illustrate it is the legend of the old man and the emperor. There are many different versions of it and most likely you've heard it before. But if not, or if you like to hear it again, here's the one set in Ambalappuzha, in the Alaphuza district of Kerala, in India. There, at the Hindu Sri Krishna temple, visiting pilgrims receive a bowl of rice payasam without having to pay for it. And the reason for this is that an old debt from the distant past is being settled, to this day. Because it was the Lord Krishna himself who, once upon a time, came to the court of the king of the region in disguise and challenged him to a game of chess. The king was a keen chess player and so of course he readily accepted the challenge and, as is the custom in such cases, asked the old man to name his prize, should he win. The old man had little use of worldly goods, being near the end of his life and receiving great happiness from wisdom alone. So he requested no more than some grains of rice. The king was taken aback: surely he, the old man, was being immodestly modest. He was a king after all, name he, so he urged him, a greater reward. But the old man declined; just grains of rice: one for the first square of their chess board, two for the second, four for the third, and so on, for each square of the board double the number of the previous one, until each square had been taken care of. The king, still somewhat wounded in his pride, relented and agreed, and thus they sat down to play their game of chess, which the old man calmly won. The king called for a bowl of rice from the pantry to pay the old man his dues. The servant brought him a bowl of dry grains of rice, and the king started putting them down on the chess board. One on the first square, two on the second, four on the third, eight on the fourth... By about square twelve, he realised he needed a bigger bowl. His servant had already counted out more than 4,000 grains of rice and clearly there were a few more squares to go. So men from the kitchen brought him a sack of it, and another, just in case. They weren't enough. Less than a third into the board, the king's staff – all of them: they had all been relieved of their other duties – had counted out over a million grains of rice, and there was no end in sight. It was probably around about halfway point, square 32, that the royal rice stores were depleted, as just over 4 billion grains of rice had been made over to the old man (it is doubtful they were still counting them individually by this stage). Clearly, though, that wasn't going to satisfy him, since half a chess board was still 'empty', as it were. By square forty, the king owed the old man in excess of a trillion grains of rice. (Assuming, for the moment, that a kilogram has about 50,000 grains, that would amount to around ten thousand tons of rice.) By the 64th square, the old man was entitled to no fewer than 18,446,744,073,709,551,615 (or 18.4 quintillion grains of rice), which might equate to roughly 368 billion tons of it. By comparison, India's rice production for the crop year October 2014 to September 2015 was 105 million tons,[CXXXVI] which would suggest that even employing contemporary farming methods of today, and giving *all* of today's India's rice crops over to the old man, the poor king would have taken about 3,000 years to pay him off. Mindful of coming across as unreasonable, the old man now revealed himself to be Krishna and said to the king, 'you don't have to pay me all at once. You may take your time.' And so it is that for quite a while yet, the temple at Ambalappuzha will be feeding the pilgrims in the name of the king until his debt to Lord Krishna is paid off...

And the moral of the story? If somebody tells you that a quantity is going to double at regular intervals for the foreseeable future, you'd be well advised to take notice. Currently, we're seeing a doubling of the amount of electricity generated by photovoltaic solar cells roughly every two years.[CXXXVII] Nobody knows with any certainty for how long this will continue, but what we do know is that what looks like a very flat line for a while can turn into a very steep curve all of a sudden.

With photovoltaics, energy sourcing for the first time finds itself on the same economic trajectory that information technologies have been on for fifty years: one on which the cost per unit goes down, the more units are manufactured. Currently, we are observing a price reduction per unit capacity of 30 % per year, every year, and we can expect this to continue. This is not quite on the same scale as Moore's Law, but it's the same principle and it changes the parameters of the game. Up until now, the more energy we required, the more strain we put on our resources and the more ingenious our methods for extracting resources had to become. Even as our technologies advanced, the price for energy has either remained the same or gone up. But if the method by which we source energy is to print foils and panels, of any size or shape, that become cheaper the more of them we print, then we *will* reach a point where we will have energy in abundance.

The amount of energy we covered from solar in 2010 stood at 0.5 % of our total energy consumption. This does not look impressive on any graph. Until you realise that the proportion doubles every two years: it is subject to exponential growth. The graph that doesn't seem to do much for a long time is deceptive:

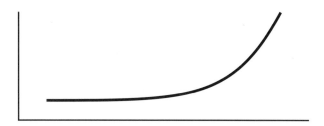

If this trend were to continue for the next 20 years or so – and this is not inconceivable – then a set of figures that look entirely unspectacular for about ten years suddenly starts to make an extraordinary impact. The corresponding table looks like this:

2010	0.5 %	barely registers
2012	1 %	still not a lot
2014	2 %	can't see it really
2016	4 %	ok, so there is some
2018	8 %	yes well...
2020	16 %	perhaps we can look at this in more detail
2022	32 %	so, that's a third
2024	64 %	two thirds then...
2026	128 %	that's ridiculous, that's everything and then some
2028	256 %	way more than enough
2030	512 %	abundance?

Which would suggest that at the current rate of increase, we could have about a third of our energy requirement covered from solar by 2022, and more than our current requirement by 2026. By 2030, we would be getting five times as much energy out of solar as we currently need overall.

But is it *realistic* to assume that this doubling of solar capacity will just continue, every year? Is it not far more likely that the curve must begin to flatten? When will it do so?

The answer, of course, is: we don't know. What we do know is that if solar cells keep getting cheaper, then there will come a point where garnering energy from the sun will be more economical than burning it from fuels or splitting it out of atoms in nuclear processors. When that is the case, we would expect the growth rate for solar to go up, rather than down, and do so even further and more rapidly than before. Also, photovoltaic solar panels, foils or cells can be part of large-scale, quasi industrial installations, but they can as easily be a consumer product: a piece of printed matter that you can order online. It is so simple and so user-friendly that, depending on what you want to do with it, you don't even have to have an engineer come round and install it for you. You can actually do it yourself. And this means that solar is subject to another extremely powerful driver: you.

You, the user, will not only determine whether or not solar establishes itself as the primary source of energy, you will also determine the pace at which it does so, you will set the conditions in which this is going to happen, and you will be one of its main beneficiaries. You'll even be able to garner it yourself and sell it to the grid and each other. And you will therefore be one of its principal investors. So if people say to you 'this is never going to happen', you don't really have to argue with them. You can just make it happen anyway.

VI   ENERGY AND ARCHITECTURE

You may remember us saying on a couple of occasions that some 40% of our primary energy use goes into buildings, principally to adjust the temperature in them.[CXXXVIII] (By comparison, only about 25% is used for transportation.)[CXXXIX] How we construct or adapt the places we spend most of our time in is therefore of considerable importance when it comes to energy, and so we want to briefly examine the role that architecture plays.

Switzerland is a country that has become very successful at putting up buildings which manage to provide very high levels of comfort while needing almost no energy at all. The talk there is of creating a '2000 Watt Society': a society in which each individual requires an average of 2000 watts of power, compared to today's roughly 5000W (which, as we've already seen, compares favourably to the 12,000 watts the average American uses).[CXL] So in direct pursuit of this ambition, houses are being built with thick walls, multiple layers of insulation and sophisticated construction materials, all to the very laudable aim of bringing down energy consumption.

Yet, by the year 2010 it had become apparent that it was actually cheaper to harvest a kilowatt of solar power in Spain than it was so save the same kilowatt in Switzerland. So in fact, architecture in Switzerland could be lighter, cheaper, simpler, freer and more flexible if, instead of having to insulate and protect against energy use locally, it could be connected to, and draw from, solar technology elsewhere, with the energy balance overall not being affected. (Which presupposes, of course, that the same kilowatt won't be needed in Spain, which is another matter...)

This makes for a thought exercise worth undertaking, because it may lead us to some surprising realisations. The realisation, for example, that thicker walls aren't necessarily better. Depending on what kind of material they're made of and what kind of uses we want to put them to, walls could become membranes that sometimes let through a lot of air and sometimes very little. Or they could treat the air that they let through in such a way that it doesn't matter what the temperature is outside, inside it is always pleasant. Who says a building can only be heated from the inside anyway? What if it were possible to use some of the energy (such as sunshine) that warms the facade to adjust the temperature from the outside in? The mantra of yore has been 'insulate, insulate' to protect our buildings from the conditions outside and thus make them 'more energy efficient'. But that may be a ludicrous thing to do when we could achieve much better results by working with the conditions outside and started to dismantle, rather than shore up, the boundary between 'inside' and 'out'. So the bulwarks against the elements, made of stone, concrete and insulation, could possibly turn into skins that work with, instead of against, the weather. And the driving forces for this are not theory or aesthetics, but economics: if it is more economical to build with thermodynamics – the ways in which energy behaves as part of a system – in mind, then chances are people will want to do so.

We will come to thermodynamics in just a moment, but before we do, we want to spin this particular thought one step further and touch, very briefly, also on architectural style; not so much as an aesthetic expression, but in the context of energy, resources, necessity, space and abundance.

The question that presents itself is this: how can we build in such a way that our dwellings, our homes, our places of work and leisure become, in effect, *applications*. This, as you will have noted, lies at the core of our thesis: that we can and really want to move from a *resources* based understanding of energy – which by definition ultimately implies scarcity – to one that is based on the abundant availability of energy and its *application*.

How, in other words, can the principle we see at work in information and mobile phone technology – where we no longer think in terms of supply and demand but take supply for granted and think in terms of the potentiality of what we can do with it – how can the principle we see at work in yoghurt logistics, where we don't think in terms of meeting our baseline requirement for dairy products, but instead think in terms of the multitude of differentiations that their ready availability offers us; how can these principles be applied to architecture.

It means looking at architecture now too as an urban laboratory for creating environments for living, for creating and managing microclimates, for integrating and enmeshing with nature, and for structuring and assimilating our actions and rhythms in tune with the world around us. Architecture then, too, moves onto a different plateau. It is no longer about sheltering and protecting us from the world as a hostile environment on the one hand and fulfilling utilitarian functions on the other, but instead it is about the wealth of potentiality we have for shaping our existence on this planet.

Like so many things, architecture right now is in a particular state of flux. If it is at all possible to talk of a contemporary architectural style, then it could be said that the crystalline, linear, edged forms of modernity have been succeeded by today's more organic forms. Buildings by Zaha Hadid,[CXLI] for example, or Frank Gehry,[CXLII] reflect the structures of floral growth: like plants, these edifices grow 'from the inside out'. In parallel to this, we also find surfaces that do in fact resemble membranes, rather than walls of brick and mortar, envelop otherwise still largely traditional types of architectural layout and construction. This can be seen, for example, in the shapes, lines and materials used by Rem Koolhaas:[CXLIII] like animal organisms of the next evolutionary level, they grow 'from the outside in'. Rather than moving away from each other, creating ever more complex but immovable objects, the various elements and structures of the building move *towards* each other, forming a differentiated and adaptable mesh.[CXLIV]

We don't want to go much further here into style and form, as ours is a book about energy, and the point we want to make really concerns not the aesthetics of architecture but the underlying approach and workings of it, because this is the first time that we've been in a situation where architecture allows for qualities not to be set out in advance, but to be constantly put in question and newly shaped. And this, in turn, is of interest because it is, in a sense, the architectural answer to a world in which a constant energy stream provides surplus quantities.

Traditional architecture was perfectly adequate for dealing with a traditional energy structure; an energy structure which was defined by the supposed stability and security that comes from fully quantifiable and delineated energy capacities: fuels. With fuels you know how much energy you've got, where it is, when it's going to turn from one thing – oil, for example – into another – such as heat – and you know how to get from one point to the other. In the new energy landscape, as we have seen, such certainties go straight out of the window. Energy ceases to be tangible, and instead of knowing how much of it sits where and what the mechanical and industrial processes are to get it from A to B, and once it's at B from one state – such as matter – into another – again, such as heat – we now know only that it's there, it's always there, and we have to connect it so it can also be everywhere else. The certainties of architecture, like the methods of energy provision, are thus being brought into question.

VII   THE ROLE OF THERMODYNAMICS

There are many ways to deploy energy, and all are not equal. It would not make sense, for example, to simply replace gas and oil heating systems with electric central heating systems and continue to treat electricity as just another 'fuel'. That is the approach that treats energy as a matter of cause and effect: it's cold, so I need to heat the house. I need to heat the house, so I burn some fuel to make the radiators hot. Now it's warm, but I use all this energy, which I have to get from somewhere and pay for. And when I've used it it's gone. That's the old way.

The new way looks at the entire system and addresses not the energy itself but the effect that is desired, and then puts into place the most efficient method to achieve that effect. So in the house, the desired effect may be a comfortable temperature, no matter what the weather is doing outside. Using a combination of state-of-the-art building materials, a heat pump to store energy in the ground for short periods, and where necessary electric heating or cooling respectively, the 'intelligent' house regulates its own indoor environment according to the setting we choose for it, by the most sensible means.

This is very different to what we've been doing so far. Instead of looking at cause (it's cold) and effect (I burn some fuel), we start looking at states or conditions: at balances and relationships. A house may be exposed to a lot of sunshine during the day and get very hot, but during the night it may get very cold. Switzerland once again serves as a fine example, though you'll find similar conditions all around the world. Here, if you go up into the Alps a bit, and you don't have to go all that far, say to around 1700 metres altitude, which is about where the tree line is in Switzerland, you can happily sit outside your chalet or hut on a lovely October day and the thermometer will tell you it's about 22 to 24 degrees. If you're in the sun, it will feel quite a bit hotter, so hot that you need to wear sun screen, a sun hat and take off your shirt. But once the sun sets, within minutes you'll want to put your shirt back on, and a jumper, followed very soon by a nice warm jacket. Give it another half hour, you'll want to go inside. By eight o'clock in the evening, you'll want the heating or a fire on, and come midnight, you shouldn't be surprised if the thermometer has dropped below zero. The traditional, cause-and-effect way of dealing with this kind of situation is to close shutters and windows during the day to keep the heat out, then open them briefly to get some fresh air in and then whack on the heating or build a fire in the fire place at night.

Now, aside from the fact that the fire crackling in the fire place is very homely and nice, none of this actually makes all that much sense. Would it not be far more efficient and more economical, if all that excess energy that we had during the day could be stored in the ground until the evening when it would be used to keep the place cosy and warm. And of course it can, it just requires a different approach.

And that approach is to put thermodynamics into practice: it's looking at a house or a building and its surroundings as a system, and finding ways of making use of the energy that is available within that system at any given time, and working out what needs to be done in the most sensible way so the people living or staying in the house have the best possible and most agreeable conditions inside. Using this type of approach, the proportion of energy that is used for adjusting the climate in indoor environments (mainly a case of making them warm and cooling them down) can be reduced by some 80 % without causing any reduction in comfort or living standards. Rather the opposite: living standards and quality of life *improve* in the process, because our creature comforts are more comprehensively and much more subtly catered for. Architects and energy engineers understand this principle and are increasingly working with it. The change, here too, is underway.

So we are moving in the right direction already. But we want to go yet another step further and also understand it conceptually, because doing so frees us from having to keep looking at energy in a utilitarian way, as something that we use, let alone use up. We can take our understanding of thermodynamics to the level where we realise that we are indeed embedded in energy and that our task is not even to 'manage our energy use', but to *organise ourselves* within the energy that is there. We have known since the 19th century that this is the case, that energy can not be created nor can it be used up, yet still for the next two hundred years we have behaved exactly as if it could.

And this is why we are going to now take a moment and look in detail at what exactly our scientific understanding is of energy, because it tells us a whole lot about what we can and can't do. If you already know everything there is to know about thermodynamics, feel free to join us again later.

THE LAWS OF THERMODYNAMICS   Thermodynamics, in a nutshell, is the study of the relationship between thermal energy (heat) and other forms of energy, such as motion or 'work'. Heat and motion are very obvious forms of energy to us, but matter, too, is a form of energy, as are radiation, sound or light. As Einstein was to observe half a century after the Laws of Thermodynamics were formulated, everything in the known universe is one form of energy or another, and so the principles that lie at the basis of thermodynamics affect everything else and are therefore of fundamental importance and relevance to our understanding of what energy is and does.

At the core of this are the four Laws of Thermodynamics, which in turn are based on the concept of thermodynamic systems.

THERMODYNAMIC SYSTEMS   A thermodynamic system can be anything from a microscopic organism, like a bacteria, to a planet circling around a sun. A system is simply whatever happens to be under observation, and is thus distinct from the rest of everything. You are a system, as is your mum. And so is a steam train.

Irrespective of what the system is, there are ordinarily the system itself, a system boundary and the system's surroundings. So if you are looking at a glass of water on a table in the room and you are interested in how the temperature of the water in the glass behaves in relation to the temperature in the room, then the room, the glass and the water together form a system, whereas everything outside the room would be the surroundings. The boundary, in this case, would typically be a wall. You may, on the other hand, only really be interested in how the water in the glass behaves if you put some ice cubes in it and consider the room irrelevant to your observation. In this case, the system now consists of the water and the ice cubes, whereby the glass now becomes the boundary and the room, as well as everything outwith, becomes the surroundings. Similarly, you may want to study what happens when you put a lid on the glass, make a hole in the lid, draw a tube from the lid and point it to a small windmill, and then start heating up the water with a Bunsen burner. Assuming the glass you're using is fireproof, you may note that the little wheel will start to turn after a while: your system now includes the burner, the flame, the glass, the water, any remaining ice from the ice cubes, the air above the water, the steam that the water turns into, the air and the steam that together travel through the tube, and the wheel. You get the gist.

There are, in thermodynamics, three major types of systems:

ISOLATED SYSTEMS   These are systems that do not interact with their environment or with any other system in any way whatsoever. Isolated systems only exist in theory, since in practice it is impossible to completely isolate a system from *everything* around it *completely.* Say you take your glass of water, put a new lid on it that hasn't got a hole in it, and put it in a perfectly airtight platinum container, which you wrap in layers of insulation and put in a concrete safe with walls three feet thick and put that safe into a vault which lies deep inside a mountain where you let it rest and never move or touch it again.

Over time, the water in the glass still adjusts to the temperature in the container, which, no matter how well you insulate it, adjusts to the temperature in the safe, which adjusts to the temperature in the vault, which adjusts to the temperature deep inside the mountain. The system is perhaps insulated well, but it is not isolated. But for practical purposes, it may be possible to have a system that is isolated *enough*, because the amount of interaction that can take place over the given period of observation is so minimal that it doesn't register as significant for the purposes you have in mind.

You could wonder, quite conceivably, what happens when you put a pear into a vacuum and leave it there. Your laboratory may be quite capable of keeping the glass tube you've put your pear into at a steady temperature and you can also seal it from light, and unless you happen to be finding yourself on a tectonic fault line or at the top of a particularly tall tower or on a boat, you can assume that your system (the pear) is not going to move or interact with anything much and is therefore, to all intents and purposes, 'isolated'.

CLOSED SYSTEMS   A closed system interacts with its surroundings or with another system through energy as heat and/or work, but not through matter. Our glass of water with its solid lid on is a closed system: it will adjust to the temperature of its surroundings over time (the room) but no water and no air and no steam emerges from it, as long as the lid is tight.

OPEN SYSTEMS   Correspondingly, an open system interacts with its surroundings or with another system through energy as heat and/ or work and/or matter. Once we add the tube to our lid and have steam travel through it to blow onto a small turbine, our system therefore becomes an open one, since steam is matter, and not just energy.

While it is not possible in the real world as we know it to have a genuinely isolated system, it is possible to imagine one, and it is therefore possible to imagine a thermodynamic system in complete *equilibrium*. A system is in equilibrium when all the thermodynamic interactions within it have taken place and therefore no further exchange of energy, be it motion or temperature, nor matter, is possible. Any system left alone in isolation would eventually reach equilibrium, but since no system can be left alone completely in isolation, no system really ever reaches complete equilibrium, other than possibly the universe itself, but that is far from certain, since we have no way of knowing, right now, what, if anything, there is outside the universe and what, if anything, our universe may therefore be interacting with.

THE ZEROTH LAW OF THERMODYNAMICS   The numbering of the Laws of Thermodynamics is a touch idiosyncratic, because it occurred to scientists only after they had written down the principles that govern the first three laws that there was in actual fact a fourth principle, which was really more fundamental than all the others and should therefore be considered first. Having so rashly labelled laws one through three already, it was decided to name this The Zeroth Law of Thermodynamics: not, perhaps, the most elegant of phrases (and a bummer to pronounce if you have a bit of a lisp), but seeing that this law itself could be reproved for stating the blatantly obvious, the name is perhaps not entirely inappropriate:

If a system A is in thermal equilibrium with a system B, and system B is in thermal equilibrium with a system C, then system A and system C are also in thermal equilibrium with each other.

What it means is this: if you take a can of ice cold Weissbier from the fridge and put it on the table in front of you, then over time the beer will acquire 'room temperature', it will be exactly as warm as the air in the room. It is then in thermal equilibrium with the room. Also in your room, perhaps lying next to you on the sofa, still on its carton, may be a half-eaten meat feast pizza that your favourite delivery boy Jason brought to you about an hour ago. As you'll know only too well from experience, over that hour or so, the pizza, too, will have acquired room temperature. Now if you consider your glass of water to be system A, the room to be system B and the pizza system C, then you can now safely assume that by the time the beer has reached room temperature and the pizza has done so also, the pizza and the beer are the same temperature as each other and may therefore have lost all their allure. (The same may not be true of Jason, who, incidentally, as long as he's alive and his heart is beating, is not going to acquire room temperature, no matter how long you leave him lying around on the sofa...)

THE FIRST LAW OF THERMODYNAMICS   Considering how much we tend to fret about 'wasting' energy and 'saving' energy, the first principal statement that can be made about energy may come as a bit of a surprise:

Energy can be neither created nor can it be destroyed, it can only be transformed.

It means that we can never hope to actually 'make' energy, nor are we ever really getting rid of any, all we're ever doing is putting it from one state into another. Since energy can come in the form of heat, motion or matter, to name but these, there's a fair amount of scope for transforming though. Every time you hop into your new convertible to visit your grandmother 25 miles out of town, you're transforming about a gallon of petrol, there and back. You're transforming it on the one hand into motion, of course, which is the point of the exercise and may be labelled 'work', but you're also transforming it into heat. The amount of energy that is in the gallon of petrol you're burning is exactly equal to the amount of energy that comes out as heat and work (plus whatever is left in sundry gases and particles, in other words leftover matter).

The problem to us arises from the fact that once the energy has done its work, it may not be in a position to do any more. So if grandma lives up on a hill, quite a bit of the energy that moves you to her will go into putting you and your wheels on a higher ground than you started off from, and so can be used again to come down the hill. But if it's pretty flat round where you live, then the energy you've used to get to her has now done its work and cannot do any more, it is therefore, to all intents and purposes, 'spent'. The overall amount of energy though is and remains the same: there is, within our universe, a finite amount of energy. It's a substantial amount, and the fact that it's finite is not cause for immediate concern, but unless something exceptional happens, for example another Big Bang, or our universe collides with another universe, or in some other way interacts and therefore gets energy from one, we can assume that the energy in our universe will always remain the same.

THE SECOND LAW OF THERMODYNAMICS   The second principle at the basis of everything introduces the concept of *entropy*. Rudolf Clausius,[CXLV] a German physicist and mathematician first used the term to describe energy that is 'already converted'. This is energy that we would consider 'used' or 'spent', as in the example above. It means that the energy was previously in a state from which, during the process of conversion into another state, it could carry out 'useful work', and has now undergone this process of conversion into a state from which it can no longer be converted into another state, and therefore do no more 'useful work', without any sort of input from the outside.

Although this may sound wilfully complicated, it is in fact tremendously simple: when you take a log of wood, for example, and burn it, then the energy in the wood converts, as you will find to your delight every time you light a fire, into heat. The heat can be just that and may not have to do anything else for you to be content, but as everybody knows, it can also be put to use: it can heat up water, turn the water into steam and power a turbine. In a system that consists of the wood, the water, the tank with the water in it, the steam and the turbine, you generally start off with a situation where a lot of the energy is in a state (such as wood) from which, during its conversion into some other state (heat) it can do 'useful work' (power the turbine), and you end up with a situation where practically none of the energy is in a state where it can do any more useful work, because it has already been converted.

In a thermodynamic system, a high level of entropy suggests that a large proportion of the energy present can *not* do any work, whereas a low level of entropy suggests the opposite, that there is a large proportion of energy present that *can* do work. Expressed a little differently: high entropy equals little work can be done; whereas low entropy equals a lot of work can be done with the energy that resides in the system.

The phenomenon of entropy stems from the fact that every thermodynamic system – whether it is an apple, a glass of water, your grandmother or the Forth Bridge – is made up of extremely small particles, such as atoms and molecules, which at any given point are in a greater or lesser state of 'disorder'. While the system – the apple, glass of water, your grandmother or the Forth Bridge – is in its 'normal' form, it tends to be highly structured and thus 'ordered'. The molecules and atoms are in a constellation specific to the nature of the system, and as it is energy that has turned the system into what it is – an apple, a glass of water, your grandmother or the Forth Bridge – the energy therein contained is separate and clearly distinct from the energy in any other system.

Depending on *how* exactly the system is structured, the amount of energy contained within it that has the potential to carry out useful work can be higher or lower. A log of wood, as we've seen, happens to be structured in such a way that it contains a high proportion of energy that can carry out 'useful work'. A slab of concrete, by contrast, really doesn't. In a log of wood, then, the level of entropy is correspondingly low, whereas in a slab of concrete it is high. What makes it so is the molecular structure of any given system: a system that has a high level of 'order' has a lot of energy in it that can carry out 'useful work', a system that has a low level of 'order', and therefore a high level of 'disorder' has little energy that can carry out 'useful work'. And when we say "can carry out 'useful work'" we always mean in the process of conversion from one state into another.

Entropy is therefore not only a measure of how much of the energy contained in a system can carry out 'useful work', it is also a measure of how much molecular *disorder* there is in a system: a high level of entropy expresses a high level of disorder, a low level of entropy expresses a low level of disorder.

Now, the Second Law of Thermodynamics states that:

> In an isolated thermodynamic system, the level of entropy always increases over time.

Notwithstanding the fact that there aren't, in the 'real' world, any perfectly isolated systems, what this means is that if you leave any thermodynamic system alone for long enough, then things will start to disintegrate, and the molecules that make up the system – be it an apple, a glass of water, your grandmother or the Forth Bridge – will start to rot, evaporate, age or corrode respectively, and ultimately crumble and fall apart. This is why the Second Law of Thermodynamics also gets cited to explain death.

The reason that in an isolated system the level of entropy increases over time is that in the microcosm of molecules and atoms as much as in the macrocosm of the universe, things seek to balance each other out. If you take a mug of hot chocolate for example, and you pour some cold milk into it, then the hot chocolate and the cold milk will not stay so for long: they will very quickly balance each other and become, together, lukewarm. In the context of temperature we are very used to this and see it as natural, and therefore obvious. In the context of other forms of energy though, it is less easy to observe, but nevertheless still the case, even if the process may, due to the amount of movement the molecules may be in, take a whole lot longer. Over time, in some cases over a very long time indeed, such as a few million years, the particles that make up a thermodynamic system – and again, that's any system, be it an amoeba or an oak tree or the universe itself – will balance each other out. They continue to do so for as long as they can.

What makes this particular principle seem a little counter-intuitive to the every day user is the fact that in the context of thermodynamics the terms 'order' and 'balance' stand at opposite ends of the spectrum. Instinctively, many of us would associate a high level of 'order' (and therefore a low level of chaos) with a great degree of 'balance'. The reverse, a high level of disorder or chaos would then imply a great degree of imbalance. But that's precisely not how it works; how it works in ther-modynamic terms is the opposite: as long as things, objects, organisms and people are in perfect order, they are also in perfect imbalance.

If you take your glass of water, for example, you have in it a quantity of perfect water molecules in their liquid state. Say your glass is half full and you now plonk into the water the exact same quantity of ice: you have just added a bunch of perfect water molecules in their frozen or solid state. At first glance this may strike us as perfectly balanced, because the glass is half filled with liquid water and half filled with solid ice. In thermodynamic terms, however, this represents a perfect imbalance, as all the solid molecules are contained within some parts of the system and all the liquid molecules are contained within the other parts of the system. What the system will do though, and you have seen this in action many a time, is create a perfect balance: it will take only a short while until all the ice in the glass has melted, the ice and the water acquired the same temperature, and so the molecules within the system have become all of the same type. As far as thermodynamics are concerned, *now* we have perfect balance, and we also have, within the system, perfect chaos, because there is no longer any order or distinction between different entities, the entities are all mixed up, and so mixed up that they will never now separate again until or unless something from the outside makes them do so.

This point, the point that something from the outside is required to create a new order in the system also means that if we want the entropy in one system to go down (remember that left on its own it will always only go up) the entropy in another system has to go up. In other words: in order for the contents of a glass of water to become separated again into molecules of a different nature which then makes the energy contained within the water capable of carrying out 'useful work' once more, there has to be another system elsewhere that therefore loses some of its own entropy: you may decide to put a heating element into your glass of water and heat it up until it starts to boil and turns into steam; the steam can then be used to power a little turbine, so while the water is boiling hot it contains a lot of energy that can be put to work. But that energy does not come from nowhere, it comes from the heating element. It, in turn, needs to be 'fed' with energy, and whether this is happening by an electric cable, or by some other method, it for certain means that in some other thermodynamic system energy that was previously in a state where it could be put to work is now being converted into energy that can't do that work any more. All the thermodynamic systems of the universe will eventually balance each other out, which is why the overall level of entropy in the universe can only go up.

One of the consequences of the Second Law, incidentally, and not an unimportant one, is that heat doesn't flow from a colder system to a hotter system, without any work input, but only the other way round. When you are putting ice in your drink, you're really heating up ice cubes, not cooling down gin. The effect, to the guests at your party, is more or less the same, and you may therefore not care too much about the difference, but in many other practical applications it is of significance: your fridge, for example, chills because it has a compressor built into it, which applies motion ('work') to temperature and takes the warmth out of the enclosed space you keep your milk in. If you want to heat up your milk, you don't need any 'work', all you do is pour it into a little saucepan and put that on the hob. The heat will move readily from the hot plate or gas flames of your cooker through the saucepan into the milk and you're ready for your hot chocolate in no time at all.

THE THIRD LAW OF THERMODYNAMICS   The Third Law of Thermodynamics is no longer that relevant to our discussion, concerning itself, as it does, with what happens to a system when it approaches absolute zero: 0 K or -273.15° C, or -459.67° F. This is a theoretical point, since it is not possible to actually reach absolute zero, even though it's possible to come quite close to it. The lowest temperatures artificially created by human beings to date lie around the femto-kelvin mark. A femto-kelvin is one quadrillionth of a kelvin. Absolute zero being 0 K, one quadrillionth, or 10-15 K or 0.000 000 000 000 001 K is really very cold indeed. Around about that point, the Second Law of Thermodynamics somewhat inelegantly goes out of the window and makes room for the Third Law, which states that:

As a thermodynamic system approaches absolute zero, all processes cease and the entropy of the system approaches zero too.

Before then, we recall, the entropy of the system had been *increasing* over time. But at absolute zero, there is zero molecular activity and the system is considered to be in its *ground state.* Since there is no molecular activity, the system can't interact with anything, either within itself or without. The level to which energy can be exchanged or converted therefore becomes irrelevant, as it no longer applies to anything we're looking at. So while each component within the system may have maximum entropy, the measure for entropy now becomes obsolete and our perspective therefore flips: we're now looking at the same thing from the opposite direction, where the two extremes meet: maximum entropy switches to minimum entropy and all is at once perfect chaos and perfect order. As this is a theoretical state only, it is not one we need to dwell on too much here, but it raises any number of really quite fascinating philosophical questions about the nature of the universe we inhabit.

VIII   UNTAPPED ENERGY: EXERGY

Thermodynamics and the 'thinking in terms' thereof offers up one more short detour we should allow ourselves before we head back onto the more obvious trail of our exploration.

And that is the concept of *exergy.* It is of relevance to us, because in contemporary architecture and energy technology, we are seeing a move towards, and considerable potential in, actively employing the laws of thermodynamics to the end of making better – you could say more efficient – use of energy that's 'already there'. One of the examples we've given cited the temperature difference between night and day and the possibility, inherent in such a situation, of garnering some of the energy (heat, light) that is there during the day and storing it up for later at night when it is no longer there, but in fact more in demand. One of the ways to do that is by, for example, using a heat pump to quite literally pump the energy from the roof and/or walls into the ground, store it there, and then bring it back up again when it's needed. Similarly, it is of course possible to use geothermal energy that already sits in the ground and bring it into the building when it is useful to do so. In each case, energy is transferred between the house and either the atmosphere outside or the ground underneath, whereby, in thermodynamic terms, the house is the system and the atmosphere or the ground are the system's surroundings. (The system boundary would again consist of the walls, windows and roof of the house.)

As we have seen when looking at the First Law of Thermodynamics, energy can not be created or destroyed, but as we've also seen, not all energy is in a state where it can do 'useful work'. Over time and without external input, a thermodynamic system will seek and invariably reach equilibrium with its surroundings, and when that equilibrium is reached, no more energy exchange can happen, the system is considered to be perfectly balanced and the level of entropy reaches its maximum. In a system that isn't in perfect equilibrium – and in practice that includes every system there is, even if in theory it doesn't – there is always some 'useful work' that can potentially be done.

Now, the term *exergy* describes the greatest quantity of 'useful work' that is possible during the process of a system being brought into equilibrium with its surroundings if the surroundings are a 'heat reservoir'. Although somewhat misleading, 'heat reservoir' is a technical term that we can't change without causing a great deal of confusion and some considerable consternation among the scientific community. It also gets referred to as a 'thermal reservoir' or a 'heat bath', and whatever we call it, it is nothing like what any of these names suggest. Because all these names suggest a pool or a pond or a tub or tank with something in it that's hot. Like a water reservoir, only hotter. This is not the case. A 'heat reservoir' in thermodynamics is a system whose 'heat capacity' is sufficiently large for its own temperature to remain unaffected when it interacts with another thermodynamic system.

Say you get home after work on a fine summer evening and it's been really hot all day, and now the temperature outside is a lovely 22 degrees but inside your house, it's got hot and you need to open the window immediately to get some fresh air in and also to get your room temperature down from 32 to something more like 24 degrees. Well in this case that is all you expect to happen. You would not expect the temperature in your town to go up by any measurable amount, let alone by any significant amount, just because you open your windows. The thermodynamic system that is your town is a 'heat reservoir' in relation to your house, because while the 'heat capacity' of your house is relatively small, so small that all it takes to bring down the temperature inside it is to open some windows, the 'heat capacity' in your town is very large, so large that you opening your windows has no significant impact on it. Of course, it has *some* impact, it's just that the impact is incredibly small and not sustained enough to register. But if you multiply your behaviour by a large enough number of residents in your town, all doing the exact same thing at the exact same time, then you will end up, eventually, with a measurable impact. Which is one of the reasons why a city like London, for example, during winter, is almost consistently about 2 degrees less cold than the countryside around it.

Now in the example of your house and your town, the constellation is such that your house is a thermodynamic system and the town is its surroundings. This is obvious not because that's how, in any case, you experience it, but because the house is defined by its boundaries, the walls, windows and roof, and it sits in a much bigger system, your town. In a situation like this, we can now start talking of *exergy*, because here we have one thermodynamic system – your house – that is surrounded by another thermodynamic system – your town – whereby the latter is a 'heat reservoir'. The *exergy* is the amount of 'useful work' that is possible during the process of you bringing your house into thermodynamic equilibrium with its surroundings, namely your town.

We can now extend this further and say: if the thermodynamic system we're looking at is not the house and the town that the house is in, but, say, the town and the countryside around it, then the 'equation' goes up directly in step, as you would expect. Now the exergy is the amount of 'useful work' that is possible in the process of bringing the town into thermodynamic equilibrium with its surrounding countryside. Again, you would not anticipate the surrounding countryside of your town to be noticeably affected by the fact alone that everybody in town opens their window, for example.

Exergy is therefore described as a 'combination state property', because the amount of exergy there is depends on the state of a thermodynamic system and its *surroundings*. This, again, makes sense: if you come home on a cold December night, you are not going to want to open the windows in your house to get a nice 24 degrees inside. You'll want to do the opposite and shut that window which has, unfortunately, been left ajar all day. And as you might expect, exergy is directly and inversely proportional to the level of entropy in your system: if the entropy is high, that means that the equilibrium is high which means there is little energy that is left that can do any 'useful work', which is another way of saying that the exergy level is low.

Of course, the town is only one part of the surroundings of a house, and not even the most relevant one. Similarly, the countryside is only one part of the surroundings of the town. For us, *all* the surroundings of a building are of interest, and that includes the air outside, the ground on which we build, and whatever there is around the building, be that a garden, a street, a village, a town or a humungous city.

But why is it of interest? Because it ties directly into this different approach to energy that we propose. Instead of just saying, 'it's freezing cold outside, I need to heat the house immediately!' and whacking on the central heating to effectively fight the temperature in the house, which has come about as a result of the temperature in its surroundings, and in the process using a whole lot of energy, most likely by converting matter (like gas or oil) into heat, I can look at the house as a thermodynamic system and consider the differentials there are between it and its surroundings at any given time. I may then find that while in the evening I do really need to heat the house and thus move the house (my system) *away* from an equilibrium with its surroundings, at other times, for example during the day, the opposite may be the case. Remember, exergy is the amount of 'useful work' that is possible in the process of bringing the thermodynamic system (the house) into equilibrium with its surroundings. We tend to ignore this. We tend to focus on the energy we need to get the house *out of* equilibrium with its surroundings: when it's cold outside, we don't want the house to be cold too, we want it to be nice and warm. When it's hot outside, we don't want the house to be hot and sticky, we want it to be cool and pleasant. So our entire take on energy is informed really by working *against* the surroundings. We have little or no interest in bringing our house as a 'system' into thermodynamic equilibrium with its surroundings, because the only time when we're happy for it to be so is when it's nice and mild outside, and then we take it for granted.

But there is 'useful work' that can be done, sitting in the constellation. And not just when we recognise it as obviously agreeable, but also when we perhaps don't expect it to come in handy. Depending on where in the world you are, the surroundings – *any* surroundings – keep changing temperature. In some places, as we've already mentioned, these changes can be quite extreme, in other places less so. But that means that our buildings keep getting *into* and *out of* equilibrium with their surroundings. For equilibrium you can here just read 'same temperature'. This means that *whenever* a building cools down or heats up as a result of the temperature around it being lower or higher than its own, there is *useful work* that can be done. What do we do with this? By and large, nothing. Up until now, very few buildings take advantage of that. But 'useful work' can be anything, it could be charging a battery, it could be powering some machinery, it could be filling a temporary energy store, it could be any number of things that are either immediately useful or that can simply be reversed when the energy is needed again, maybe not as work but as, for example, heat.

So working with exergy means, at a very fundamental level, understanding energy in a much more comprehensive and systemic way and, again, acknowledging that the energy is in fact just there. It's not a case of generating or producing it, it's a case of making it useful.[CXLVI]

IX   INFINITE SOURCES OF ENERGY

We have laid down various strands of thought and taken a fairly detailed look at areas that are not so much tangential as perhaps backdrop to the 'energy question'. Against this backdrop and given this context, we don't want to lose sight though of what it is that we are now able to do:
- Use resources for our material, not for our energy needs
- Use infinite sources like solar, wind, water, tidal and geothermal for our energy needs
- Network electricity

All of this can happen immediately. In many places, we've already started to do this, and there is no reason why the process would have to last more than about 25 to 30 years to complete. Also, it can happen smoothly, without any major interruptions to our economic and cultural development, and it can happen across all layers, meaning it is applicable not only to the developed, post-industrial world, but just as much to the developing world.

We have spent a fair amount of time talking about networks and the point of networking electricity, we've elaborated on an approach to energy that doesn't rely on resources, thus freeing up resources to be used for the manufacture of products and then recycled, but we haven't really gone much into any kind of what used to be thought of as energy generation, but what is really a harnessing of energy, other than solar.

As we have explained, we see solar, particularly photovoltaic solar, as a strong contender for becoming one of our primary methods of energy garnering over the next few years. Because of its qualities, its flexibility, its versatility, its extremely low entry threshold, the fact that it doesn't involve any mechanical or industrial processes other than in the manufacture of solar panels or foils, and because this manufacture is becoming cheaper every year while output is increasing exponentially, we expect electricity from photovoltaic solar to be more economical soon than electricity generated from traditional power plant. And as part of a dynamic energy network, we see a shift happening away from traditional power generation to energy trading and sharing that is much more similar to how we trade and share information today.

But we don't think that we either need or want to do away with other methods of harnessing the abundant energy we have. Many people would argue that we can't *afford* not to use them. We prefer to think of this as a matter of choice rather than of necessity, but in any case, these methods merit looking into, albeit here briefly. We are not going to elaborate in much depth here, because not only is technological progress rapid in all these areas and will render much of what we're about to say out of date very quickly, but also bookstores, libraries and the internet are stocked to the hilt with literature on these subjects, and you will find a great wealth of information on all of them within seconds of entering the term 'energy' into your favourite search engine.

Nevertheless, all of the following methods are being put into practice now, and in each case with some promise. They may be broadly divided into 'mechanical' and 'thermal' ways of harnessing the elements, with the exception of photovoltaics, which we've already dealt with, and the production of solar fuels, which uses solar power to make combustible fuels by a process of electrolysis.

Here then is a synoptic overview:

MECHANICAL TECHNOLOGIES

During a conversation we were having with an elderly gentleman about harvesting the weather, he was quick to retort, 'I don't want any of those windmills ruining the landscape'. Other people consider them a thing of beauty. Wherever you stand on wind farms, they are perhaps the most visible and as a result maybe one of the more controversial ways of harvesting the weather. They are also a prime example of the mechanical 'class' of methods of doing so. But they are certainly not the only one; others include tidal, wave and traditional hydropower. Harnessing the power of waves and the tides is relatively new, and therefore the technology here may be considered to be in its infancy still. Traditional hydropower and wind though have been around for as long as we have known how to make use of electricity, and as purely mechanical machinery for centuries before then.

None of them will ever run out, so they are all genuinely infinite sources of energy, but because they use moving parts and mechanical processes, in many cases exposed to extreme conditions, they all require a fair degree of maintenance, expertise and engineering.

WIND   There are several different types of wind turbines that turn the natural flow of air in the atmosphere into electricity. The principle by which they operate is more or less the same for all of them: wind propels a rotor and the resulting motion is converted into electricity. It's straightforward and simple, even if the challenges associated with it can be formidable. For example, the amount of energy that a wind turbine generates increases exponentially with the strength of the wind it is exposed to. So in terms of efficiency and yield, it would appear to make sense to place wind turbines in the windiest locations available. But of course the stronger the winds, the more liable they are to cause damage, and the stronger therefore the materials have to be for a large wind turbine to withstand the elements.

State-of-the art wind turbines with blade spans of up to 90 metres are mini power stations each in their own right, which automatically turn themselves in the direction of the wind, angle their blades, and shut themselves off as required at wind speeds of about 90 miles per hour and above.

Apart from the aesthetic argument, complaints about their noise and the possible danger they may pose to birds, the greatest drawback of wind turbines is that the amount of electricity they produce varies enormously and is extremely difficult to predict even in the short term. Even so, there are 'good news stories' from places as diverse as Denmark, which generates some 40 % of its total electricity requirement from wind;[CXLVII] Spain, Germany, Great Britain, where the London Array, the world's largest off-shore wind farm, went into operation in 2014; to India, which is home to one of the most successful wind turbine manufacturers; China, which in 2015 reached a generating capacity of 145 gigawatts from wind (roughly the equivalent of 150 average size coal power stations); and Texas, a US state which has been associated with nothing so much as oil for the best part of a century and which now boasts some of the most ambitious wind farming projects anywhere in the world.

Harnessing the wind on a small scale, say with the occasional wind turbine on your roof or in your garden or on your farm land, is considerably less efficient purely at a return-on-investment level, but need not be dismissed roundly as folly. Also, in remote areas that are not connected to a national or international grid, individual small turbines, particularly when augmented by solar power, can provide sustained, independent energy that requires comparatively low maintenance and is, the original investment apart, practically free.[CXLVIII]

WAVES   Waves are really wind power converted into the motion of water. And wind is really solar power converted into the motion of air. But despite the fact that waves are therefore really 'solar power by two degrees of separation', they can still yield remarkable quantities of energy. How much of this can actually be harnessed, let alone harnessed economically, is another matter, but even if it's a small proportion, it may still make a vast difference, most particularly of course to countries with coastlines that have 'busy' waves. Pioneering wave farms in Portugal and Scotland showed some promise, with further development happening in the UK in Scotland and Cornwall. Australia, meanwhile, stands at the forefront of wave and tidal power technologies, successfully connecting the world's first commercial-scale wave power farm to the grid near Perth in early 2015.[CXLIX] Carnegie Wave Energy, the company behind this particular project, believes that waves have the potential to supply up to a third of Australia's electricity requirement.[CL]

While also to some extent susceptible to changes in the weather, wave farms in most feasible locations provide a steady and dependable source of power, but it is still too early to say which systems will establish themselves and stay the course.[CLI]

TIDAL   We have, on one or two occasions used the expression 'harvesting the weather'. Tidal power does no such thing. What tidal power does is use the, in places extremely strong, water currents created by the tides, and turn them into electricity by way of large, sturdy turbines that sit on the ocean bed. And the tides, though a perfectly natural phenomenon, are not caused by the weather, but by the gravitational forces of the moon and, to a lesser extent, the sun, depending on where in their orbit earth and the moon happen to be respectively.

Britain has among the strongest tidal races in its vicinity and therefore is one of the potentially highest-yielding locations for this form of energy, along with Canada. Tidal power too is as yet in its earliest stages, with notable projects in Wales, where a 320 megawatt tidal lagoon power plant near Swansea now under construction is scheduled to form part of a six lagoon array that could generate about 8% of the UK's electricity demand,[CLII] South Korea, which has the world's largest tidal power plant with the 245 megawatt Sihwa Lake Tidal Power Station,[CLIII] the Pentland Firth in Scotland, which is expected to be the biggest tidal turbine array in Europe when it goes into power generation in 2020,[CLIV] and the Bay of Fundy in Canada,[CLV] but we suggest keeping our minds and all options open: the proof of the 'pudding', here as elsewhere, will be pretty much in the 'eating'.[CLVI]

HYDROPOWER   Hydropower stations are not a relevant option for all parts of the world: in order for them to be able to make any significant contribution you need plentiful water in either strong rivers or flowing through mountains. But wherever you have the latter, they can fulfil the purpose of storing energy, and that in itself can come in very handy. We've pointed out elsewhere that we don't really consider storage the

184

big issue in the longer term. In a fully networked energy model, it won't be necessary to store energy, other than for very short periods locally, such as in the battery of your phone or your car or in the bespoke energy store of your home. But that doesn't mean that in the mid-term and in certain specific geographical locations, hydropower could not be extremely useful.

Currently (October 2016) hydropower contributes around 16 % (some statistics have 17 %) of global electricity generation and about 85 % of global electricity generation from renewable sources, which the International Energy Agency (IEA) forecasts can double by 2050.[CLVII] While hydrodams, in particular, bring their own environmental concerns and geographical challenges, one of the main advantages of hydropower is its great reliability and the fact that it can be managed in tune with demand with great ease.[CLVIII]

THERMAL TECHNOLOGIES

To what extent the label 'thermal' here is accurate or useful is probably debatable, but what we mean by it is technologies that employ heat and temperature exchange of some description to generate energy. Two of the most relevant examples in this context are:

GEOTHERMAL   The World Energy Council estimates that up to around 8 % of global electricity demand could be met by geothermal installations, while giving the actual installed capacity at about 11 gigawatt, which is around 2 % of global installed capacity.[CLIX] Around two dozen countries use geothermal energy for the purpose of generating electricity, of which Iceland and El Salvador both cover more than a quarter of their electricity demand from geothermal. In California, USA, over 40 geothermal plants contribute around 7 % to the state's electricity demand. Overall, the United States is the biggest producer of geothermal electricity with well over 3 gigawatts installed capacity.[CLX]

Whether used directly to heat individual buildings or settlements (or even towns), or indirectly to generate electricity, the principle of geothermal energy rests on the fact that the temperature in the ground gets higher the deeper you bore. By sending a liquid down a geothermal well, you can bring some of the energy contained in the ground as heat up to the surface, where, depending on how hot the liquid is, you can use it either to heat up dwellings, offices, factories, greenhouses, pools or public buildings, or you can use it to power turbines that turn the energy into electricity.

This makes geothermal a suitable source of energy for some regions of the globe only, namely those where the geological and topographical conditions make it feasible to get at sufficiently high temperatures near enough the earth's surface.[CLXI]

CONCENTRATED SOLAR POWER   Concentrated solar power (CSP) describes energy collected from the sun by concentrating the sun's rays onto a focal point or focal line with mirrors, to heat up a liquid (normally an oil) which is then used to boil water to produce steam to power a turbine to generate electricity.

There are various different approaches to this. Probably still the most common one uses long parabolic mirrors which concentrate the sun's rays onto a tube running directly along their focal line. These mirrors can be of substantial size, spanning several metres across. Correspondingly, depending on where the installation is located, the temperatures generated can be very high, up to around 400 degrees Celsius. Not surprisingly, therefore, applied on a sufficiently large scale, this can generate a whole lot of electricity.

The same applies to a slightly different approach, which positions a large number of mirrors around a tower and reflects the sun's rays onto a focal point on that tower. There, as with parabolic mirrors, a liquid is heated up to very high temperatures, and that energy in turn gets converted into electricity via steam-driven turbines.

Because this technology is entirely reliant on intensive sunshine, it's only really useful in regions where the sun shines every day, most especially deserts. And there the potential is huge: Spain, the United States, India, the United Arab Emirates, Algeria, Egypt, Morocco, Australia and China all generate some of their electricity through CSP, with Spain and the USA accounting for the lion share of a global generating capacity of 4.4 gigawatt by 2014.[CLXII] The world's largest CSP project, however, is under construction in Dubai and scheduled to have a capacity of 1 gigawatt by 2020 (about the same as a standard size coal power station), increasing to 5 gigawatt or roughly the same as two standard nuclear power stations by 2030.[CLXIII]

Since 2008, the installed global electricity generating capacity from CSP has grown exponentially, doubling every two years,[CLXIV] and there is no doubt that there is still enormous potential for growth, especially as previously oil-rich gulf nations increase their investment in the technology.[CLXV]

SOLAR FUELS   While we've been working on this book, every so often a new headline has hit not only scientific publications but also the mainstream news, heralding significant advances in one area or another, often with excited predictions about the impact they will have.

One such area is artificial photosynthesis, but we tend to eschew this term because it covers two really quite different areas of research, in both of which enormous strides are being made.[CLXVI] One of them is the development of artificial leaves or 'super plants', which emulate what natural plants do and turn solar energy into organic matter that can then be harvested and burnt. There is considerable interest in this, but to our minds it is by some margin less promising than the other method,

which is to use solar power to create fuel, for example by turning water into hydrogen or by using a catalyst – a chemical compound that
facilitates a reaction without itself being affected by it – to break down
carbon dioxide and turn it, together with water, into a hydrocarbon.

Hydrogen is both the lightest and also most abundant chemical
element in our known universe, and while it doesn't occur on its own
very much on earth, hydrogen gas (H_2) is now used as a successful 'clean
fuel', because its only byproduct when burnt is water. Hydrocarbons,
on the other hand, are what we burn in oil, gas and petrol: they are
contained in massive quantities within our most prevalent fossil fuels.
When we burn fossils, we release the carbon that has been bound by
nature's very own photosynthesis over millions of years, and thus cause
a surplus of it in our atmosphere. Solar hydrocarbon, by contrast, is carbon neutral because it first binds the carbon, before re-releasing it. You
could, therefore, also refer to this method as a way of 'recycling' carbon.

While some of the research in this field goes back as far as the
very early 20th century – in 1902 the German scientist Hermann Emil
Fischer [CLXVII] received the Nobel Prize in Chemistry for work he carried
out in sugar synthesis – the way we understand and are beginning to use
it today is barely more than a decade old. You can, therefore, expect big
stories to emerge here for some time yet. [CLXVIII]

One of the reasons solar fuels hold so much promise is that they
are absolutely compatible with our existing, fossils-orientated infrastructure. We have gone to great lengths, in this book, heralding, and in a sense
welcoming, the end of the fossil fuel era, but we've also been reminding
ourselves periodically that this can't happen overnight, even if it probably
is going to happen much faster than most people currently are prepared
to expect. Still, we are certainly in need of bridging technologies which
allow for a smooth transition, and solar fuels can provide just that. They
may even become of lasting value: we have, over the millennia, developed
not only a dependence on, but also an affinity with, and appreciation of,
fire. We like it and so we may not want to banish it completely from our
lives. And for the purpose of storing energy and then releasing it in very
precise quantities, fuels have real, tangible advantages over other types
of energy storage, such as batteries, especially when we're on the move.

Solar fuels, therefore, give us good reason to look forward to the
coming years and decades with a degree of equanimity: we don't have
to panic and rebuild a whole global infrastructure by the end of this or
even the next decade, because there are environmentally sound alternatives to fossil fuels that fit into the existing one almost seamlessly. [CLXIX]

X   NUCLEAR

We are not, as we have pointed out before, minded to take a principle or
ideological stance for or against any type of energy. To us this is not what

the 'energy debate' is really about, it's not something we are fundamentally interested in and it's not what we think will get us much further.

But anyone making anything resembling a comprehensive contribution to this debate will have to sooner or later say what they think about nuclear power. The hapless Tom Aldous in Ian McEwan's satirical novel *Solar,* when making the case for solar power to his Chief, the fictional Nobel laureate Beard, describes the sun as 'a nuclear power station up and running with a great safety record ... nicely situated 93 million miles away.' And he dismisses nuclear with three words only: 'dirty, dangerous, expensive'.[CLXX]

All this, as far as we can tell, is borne out by experience. Nuclear is certainly expensive. Vast amounts of money have been and are still being invested in nuclear technology in a drive away from carbon fuels. With the amount of money it costs to plan, build, maintain and protect against potential adverse impact – terrorist attacks, earthquakes, tsunamis, aeroplanes set accidentally or deliberately on a rogue course – a nuclear power station stands in disadvantageous proportion to what it can produce in energy over its life cycle.[CLXXI] Since we've started working on this book, the world has witnessed its most serious nuclear disaster to date at the Fukushima plant in Japan, which got underway on 11th March 2011 with an earthquake registering 9.0 on the Richter scale, followed by a devastating tsunami. The fallout from this may well last for decades.

While it is still impossible to estimate the actual long-term human and environmental cost of the accident, it has since been assessed at the highest level, 7, on the International Nuclear Events Scale (INES), along with the Chernobyl disaster of 1986.[CLXXII] By 2011, it was estimated that cancer fatalities attributable to exposure to radiation from the accident in Chernobyl lay in the region of 4,000.[CLXXIII] So, certainly, nuclear power isn't 'safe'. And the nuclear waste problem persists. Sixty years into civil nuclear programmes, we to this date don't have anything close to a genuinely satisfactory solution for it.

Contrast this with a simple calculation you can make about solar power and you realise very quickly that the idea of banking on nuclear for economic reasons doesn't stack up:

As we write this, it costs about USD 500 million to set up, from scratch, a photovoltaic solar factory, and it takes about three years to do so.[CLXXIV] That's a lot of money, and one of the reasons why, after all this time, solar is still only just coming out of its pram, metaphorically speaking: the upfront investment is substantial, but the process itself, if the money is in place, is relatively straightforward.

By comparison, in Switzerland, for example, the planning process alone for a nuclear power plant (that's not actually building it, just getting to the point where you can start doing so) used to cost easily twice as much and took at least eight years (Switzerland, like

Germany, is no longer pursuing a nuclear programme).[CLXXV] A 500 million dollar solar panel factory can produce enough solar panels within a year of its operation to harvest between 20 % and 25 % of the output of a nuclear power station. In other words: once a photovoltaics factory has been producing foils or panels for a year, the amount of energy its output can generate when facing the sky amounts to the same as approximately a fifth to a quarter of what a nuclear power station can generate.

This would suggest that after 4 to 5 years, a photovoltaics factory will have made enough solar panels to cover the output of a nuclear power station, or, expressed differently, four or five such factories will be able to produce, within a year, the output of a nuclear power station. Four or five factories, let's say five to err on the side of caution, would thus cost about $2.5 billion and would then continue to put out the equivalent of a nuclear power station, at no extra capital investment, every year. Over the course of 25 years, not an unreasonable time scale in energy planning, you could say that the output of five standard size photovoltaics factories at a total cost of $2.5 billion will produce as much electricity as 25 nuclear power stations, therefore putting the capital investment for one nuclear power station's equivalent of solar energy at $100 million.

The Nuclear Energy Agency (NEA) estimates the cost for an average nuclear plant with two reactors to range from about USD 10 billion to USD 12 billion.[CLXXVI] So, at the most conservative estimates, for the cost of putting up a nuclear power plant that then for the next 25 years or so churns out the electricity of one nuclear power plant (itself) and a considerable amount of nuclear waste, you can have the electricity of one hundred nuclear power plants and no waste at all. (Photovoltaic solar panels and foils, you may recall, are nearly 100 % recyclable, and since they contain valuable materials, a thriving market for used panels and foils exists.)

But there is another consideration to be made, when it comes to nuclear. And that is this: do we want our energy to be something that we can handle ourselves, that we can 'own' as part of a network of informed participants big and small, or is it something that we want to be controlled by a hardware-intensive, heavy-security, super-high-specialist expert industry that has to be tightly controlled and rigidly regulated. In other words: do we want to pursue our energy policy along the old route of top-down hierarchical structures that are inflexible, immutable, impenetrable and extremely undemocratic, or do we want to pursue it along the 'open network' principle, which is the direct opposite: dynamic, adaptable, accessible and wholly democratic. All things considered, and we have considered them and found that the outcome in terms of the amount of available energy is potentially roughly the same, we emphatically opt for the latter. So yes: nuclear can also bring us plenty of energy, it is possible. But no: we don't think we should continue on that route.

One more interesting aspect to this is worth mentioning, because it once again gets at the very fundamentals of what we are talking about. The two technologies, nuclear and photovoltaic, have something in common that is really quite unexpected: they both employ the same science: quantum physics. But they use it in diametrically opposed ways. Nuclear fission splits atomic nuclei to release substantial amounts of energy, and nuclear fusion is what the sun does: nuclei are fused together to form new ones, in the process also releasing even more substantial amounts of energy. In both cases, the nuclear processes are of such effect and magnitude that they require an extraordinary amount of control, meaning that extremely heavy, expensive, complex and highly secure technology is required to steer these processes and make them usable in a human-friendly way. You cannot wander down to your local DIY store or go online to order a bit of nuclear fuel and make some energy at home. In the first instance, it wouldn't work, and even if you managed to get it to work, you'd most likely cause a disaster.

Photovoltaics, instead of breaking up or fusing together the nuclei of atoms, displace electrons in them. As we have seen, PV panels or foils are layered such that when photons hit them from the sun, they set in motion a flow of electrons, and that is what 'produces' the electricity. It is so simple and so gentle that you can easily go online, right now, and order, for about €5, a little PV cell with some fairy lights attached to it, and you can put the set out on your balcony and leave it there with the panel pointing to the sun for a few hours, and then when you sit outside in the evening, with your glass of wine, they will twinkle at you, these lights, until it's time to go to bed. The closest thing to a disaster you can cause is spilling your Pinot Noir over your shorts.

We are talking a lot about networks, about our emancipation from rigid structures, about a democratisation of the energy landscape, about control being taken out of the hands of the few and put into the say of the many, about opening up the energy market to all comers, about the way in which energy can be something that is 'just there', that we can effectively take for granted. On a quantitative level, as we've just noted, nuclear power can probably match photovoltaics. (In the short term nuclear can exceed PV by some margin, but in the mid-to longer term, PV, as we've also seen, will outperform nuclear. And there is no reason to assume that we will run out of nuclear fuels: while current uranium stocks are estimated to last for about another 200 years or so, next generation breeder reactors that recycle nuclear fuels could make nuclear power last for as far as we can think ahead.[CLXXVII]) And so what we have a choice of is between two extremes of the spectrum: high-complexity, high-security, high-risk, extreme level of authoritative control by the state and monopolist state-appointed contractors with highly specialist skills that are difficult to impart and export to, for example, poorer, developing countries, or low-complexity, zero-security, no-risk, no-control technology, with no monopolies, but specialist skills that are easy to teach and to share with people in any part of the world. This makes it sound like a fairly stark choice, and it is. But the important thing is that it *is* a choice, you do not have to believe people when they tell you there is no alternative to nuclear. There most certainly is.

A PLANET IN CRISIS: IX
INTELLECT
TO THE RESCUE

If some, any, or indeed all of what we've been talking about so far sounds like a jolly good idea, that still does leave us with some pertinent questions, such as: what's going to happen next? How long is it going to take? Who will make it happen and how? And we'll be getting to some of these in just a moment.

Before that though, we want to posit a perspective that sets out our view of the kind of planet we see emerging in the context of the kind of planet we're coming from. This can be phrased in any number of ways, and the one we want to propose here is one that looks at how we have, over time, regarded our sources of energy, as *what*. This, as you will have noticed by now, is of great significance to us, because whether we have plenty of energy or not enough, whether we have clean energy or energy that destroys our environment, whether we have, all of us, the chance and *permission*, even just from ourselves, to grow and develop depends very much on how we regard energy, and as what.

I   THREE GENERATIONS OF ENERGY

One way of looking at this is in terms of technological generations: phases or eras of development that build on each other and through which, you could say, we grow, maybe even 'grow up'.

FIRST GENERATION: THE PROPORTIONAL PLANET   A first phase stretches from our very beginnings as civilisations to about the time of the industrial revolution. During this very long period, and therefore very obviously for the vast majority of our existence as cultural beings, we see energy as something that is simply there, in the objects and phenomena around and within us. A 'given', so to speak. Wind, horsepower, our own muscle power, wood that we find and grow and then burn, the water of rivers and streams that we use to drive mill stones. There are mechanics and there are inventions; materials are worked and altered; there is design, and there is development; but all the way through, energy is something we view as inherent in the places and the things we find, more or less as we find them. The moment we take the millstone away from the mill by the stream, it will stop turning, unless we attach some mules or some other, newly local and specific and contained source of energy to it.

Our use of energy is, in a sense, 'in harmony with nature'. This is in itself an over-simplification, we are aware of that. By cutting down large areas of forest, for example, we radically alter the environment, and many a hedgehog or boar would not find the people settling on their habitat or having them for dinner to be all that in tune with their own priorities for survival, but it is fair to say that overall, our use of energy and therefore the impact we have as a species is proportional to our number and our spread.

You could argue that this era or generation is defined, in energy terms, by the resources and sources of energy that are there, unstable and finite, but, because of the proportions, sufficient. And you could also say that this is a technological generation of potency: of using energy for what it's worth, making use of it *as it is.*

SECOND GENERATION: THE BALANCED PLANET   During a second phase, starting roughly with the industrial revolution and lasting until the beginning, approximately, of the digital age towards the second half of the 20th century, we don't of course stop using energy in the above ways, but we now shift our perspective quite considerably. We now primarily start to view energy as a source of heat, and apart from the heat itself, which we still need, just as before, we now use that heat to generate motion. And we do this, as the name we've given to the period suggests, on an industrial scale. Energy is still a 'given', but we seriously manipulate it now, and we *systematically* use it to power machinery and apparatus, irrespective of where we are and what the locality provides. While before, if we wanted to drive a genuinely heavy millstone, we really needed it to be in the direct vicinity of its source of energy, such as the stream, we can now take the energy from the earth (coal) and use it to power a big steam engine which is capable of producing many times the amount of force we had by the stream. We become independent of the 'given' energy situation, and shape our own.

Our energy use is now way out of proportion with our number and our spread. The amount of energy we unleash from ancient, prehistoric energy stores, such as coal and oil, allows us to not only multiply faster and live longer, but also to much more radically alter our environment, to make things, to travel by methods, and cultivate our consumption in ways, never seen or done before. We become independent from the natural rhythms of day and night and the seasons, and instead start to think in terms of work and leisure as widely applicable concepts.

While disproportional, our use of energy is nevertheless balanced, at least for some time. What we experience today is very much the end of this era and so we see this not so much as the 'balanced planet' but as the 'unbalanced planet'. We feel that the tipping point has been reached and that the levels of pollution caused by our energy use have become critical. This, the planet becoming unbalanced, is however an expression of a 'balanced planet'. In order for it to be 'unbalanced', there has to be an equation that supposes it ought to be 'balanced'. And this is the case: for the 'balanced' planet to work, the energy that we use has to be managed in a way that doesn't upset the equilibrium of the ecosystem: 'balancing' is, therefore, a priority, and when things get out of balance, as they are now, we quickly feel the impact of this all around us. Which is also why this is a planet of limitations: if you go too far, and upset the equilibrium, then you reach the boundary and can go no further. We are, today, acutely aware of just this being the case, which is why many people feel exceptionally alarmed at this precise imbalance.

This era could otherwise also be characterised as one of resources that are finite but that give us potential, because we can release the energy in these resources and make things possible that weren't possible before.

THIRD GENERATION: THE GENIUS PLANET   The third period is the one we've only recently begun. Thanks to technology, which we have developed through the previous two eras, we now do not have to view energy as an earth-bound substance any more. Instead, we can tap into the infinite energy stream that reaches us continuously from the sun and make use of the energy that's around us, all the time. We are, in one sense, almost connecting back with the first era, when we used energy as and where it was, but we have since learnt how to make that energy abstract, and so we turn it into electricity and we network it. Energy is now everywhere and it is not used up, instead we simply 'tap into it', we become, you could say, a part of it. Because of this, neither does it have to be proportional, nor does it have to balance out: it is just there, boundless. And so, therefore, is our potential. It turns from potential to potentiality: the potential of potentials. No longer is what we can do determined by the type of energy we use; rather, the energy can be used for everything and anything, and we ourselves, by designing applications, determine what energy is good for, we *orchestrate* it.

For us today, the question of how we view energy is so interesting because we find ourselves right on the cusp of this shift from the 'second generation' to the 'third generation' of energy technology. And in human, historical terms, the 'third generation' has barely started, it's only half a century old. So no wonder it still puzzles us. For us, viewing energy as something 'abstract' or in any sense removed from substance, namely earth substance, such as coal, oil or wood, but also of course gas, something at any rate that we can burn and turn into fire before doing anything else with it, does not yet come 'naturally'. But that is just because we are so used to the 'second generation' way of looking at energy, which in turn we are because we are still a part of it, just.

In northern Tuscany, Italy, at the very end of a remote valley, far off any beaten track, set in amongst a little wood, there is, as we first write this, a blacksmith who operates his forge exactly the way his family has done for five generations. The fire in his forge has been burning without interruption for 200 years. Through the Italian *Resurgimento,* the formation of the Republic, World War I, Fascism and World War II, through the heyday of Italian cinema in the 1950s to Berlusconi's *Bunga Bunga* parties: whatever happened in the rest of Italy, nothing of significance changed here.

He's the last of his kind and his sons don't want to take over. So when he retires or dies, the tradition will end. But until then, he does what his father and forefathers have done: he keeps a big coal fire burning and when he forges his iron he flattens it. That's what he does, and that's why he's called *'Il Distendino',* because he distends the iron and shapes it into anything that's useful when flat: a knife or a shovel. Or a sign. The hammer he's using to do this is propelled by the stream that runs by his workshop. There's a set of gears that engages with the wheel that's driven by the water. When he needs hammering, he moves a big iron lever, and the hammer, which has the size of a microwave oven, bangs down on the anvil. You don't want anything that's not meant to be flat in there, because the force with which it does so is startling. And the noise too.

The way *Il Distendino* uses his stream to power his hammer is completely in tune with the setting. He never takes more energy out of the valley than the valley has to give him and he never does any damage to it. The same can not be said for his coal fire, of course. But what he's doing, *Il Distendino,* and most likely without knowing it, is maintain an approach to energy that sits entirely within our 'first generation of technology': he uses the energy effectively as it is, locally, to its primary purpose. The motion of the water is converted into the motion of the hammer, and the heat of the fire is used to soften the iron. The fact that he's using coal rather than wood makes a difference only in the amount of heat he gets. He does not change either form of energy and he doesn't interpret it or invest it with any additional significance or message.[CLXXVIII]

From *Il Distendino's* valley in northern Italy, it's about a six or seven hour drive to Switzerland. Once you get into Switzerland, it will take you another three hours or so to traverse it all the way from south to north in a more or less straight line (as straight a line as you can get in a mountainous country), veering only slightly west, until you get to Basel. Basel is an interesting city for many reasons, not all of which are entirely relevant here. But if you can, go and spend half an hour by the *Fasnachtsbrunnen,* the fountain right by the theatre. You'll see people sitting around it, mesmerised by the motion of a set of pointless machines positioned in a pool of water by the artist Jean Tinguely.[CLXXIX] 'Pointless', yet to great purpose, because they amuse. They entertain. They intrigue, they engage. People *love* the Tinguely fountain, and it's easy to see why: it's charming. It's mechanical, but it's *real* and it's wholly unthreatening. You can see how it works, and it works beautifully. So beautifully, the city authorities allow it to freeze over every year. So every year, the water jets and tubes and wheels and little bowls form magical ice sculptures, and as a result, every year the city has to employ a team of engineers to repair the damage the ice does to the fountain. But so beloved is it of the people, and of such value to the psyche of the city that, like an ancient ritual, this happens every year.[CLXXX]

The Tinguely fountain is anything but ancient though. It was commissioned as part of the then new theatre complex in the mid-1970s, at a time when we were well into the beginnings of the 'digital age'. Yet although the fountain doesn't 'do' anything, it serves as a fine illustration of how comfortable we are with mechanics, kinetics. Tinguely uses the principles and motions of the steam engine, and while he actually powers his 'machines' with electricity, the look and feel we get is exactly that of the steam engine. If you have kids, take them to any fairground or historical attraction where they have steam trains: you can bet your absolute bottom dollar that they'll be fascinated by it. Because this is something that we do understand: the motor. The powered machine, the apparatus. It starts with the steam engine, and nobody needs to explain what it does: it accelerates things, dramatically. Now mechanics take on a dimension that is both awe-inspiring and monstrous. And beautiful. Now we shift tons of metal on rails. Now we make stuff by the *million.*

And for a while, where you put electricity here is up to you, because, yes, it may well be 'abstract' energy, but at first we don't really use it as such. We use it the same way as we use wood and coal and then diesel and petrol: we use it just like any other fuel. Only since recently, as we get to grips with the third technological generation, do we actually start to use electricity to its purpose, which is not to give it a purpose at all, but let it be the power behind *any* purpose. It doesn't make any difference to the energy we are using – electricity – what we use it for. Electricity is always the same. Whether it pulls a train at breathtaking speed or slowly simmers a stew on the hob. Whether it pushes water through an espresso machine or scans a patient's body in an MRI tunnel. Whether it floodlights a football stadium or illuminates a single pixel on your computer screen. Now the energy is *anywhere.* In any quantity, to any purpose.

Take *Il Distendino's* hammer. In our scheme here, it's a first generation mechanical machine. It can do one thing and one thing only, which is to bang down on the anvil. And it can do so with the speed and the force that is set by the stream running by the forge. There is a straightforward one-to-one relationship between the machine and its purpose. But of course, as far back as the late 1800s a great Italian entrepreneur could have arrived in *Il Distendino's* valley, have bought it and set up a factory. The stream would no longer have sufficed, but he could have laid down a railway track to bring in coal and, using a big heavy steam engine, he could have had *twenty* hammers and anvils going, each operated by a team of two newly trained blacksmiths. It would be wrong, then, to suggest that such a machine as this did not have potential: it would be churning out not just shovels and knives and signs but anything made of iron, because there would be no need to restrict yourself to simply distend it. The relationship between the machine and its purpose has become one-to-many. And in the same way, it is no longer one fire that heats one room in the houses of the cities, towns and villages, and one candle that gives off one light, but it is now a big fire in a power station that produces electricity that sends heat and light to many households.

But imagine another two hundred years later somebody coming along and saying to the family: don't worry about manufacture. We can 3D print anything. What you need is a small design team with some decent computers and some good modelling software. What you have now is a completely open playing field because precisely what your designers come up with is absolutely not determined by anything other than the limits of their imagination and possibly your brief. Of course, you would be right in saying: a computer model of a knife doesn't cut the mustard, or anything else for that matter. You still need someone or some machine to then actually make the knife. And that is indeed the case. But see where in the order of things the actual object, where the process of making the idea material, comes: right at the end. Rather than saying: here is a machine that can flatten iron, so let's flatten some iron, you can say: here is a machine that can do anything. Let's see what we want to do with it and then put the rest in place. And with today's 3D printing technology we are also there in practical terms: by now it is actually possible to attach a machine to a computer, and it will be able to make you a pair of shoes, based entirely on your virtual design. Or part of an airliner. Or – there is nothing to stop you – a knife or a shovel...

With digital technology, unlike with mechanical machines, the uses we may put the technology to can not only not yet be determined, they can't even yet be *imagined.* In fact, it is the machine itself that enables us to imagine the next level of uses it may have.

And this takes us right to the heart of why it is electricity (what we refer to as an 'abstraction' of energy) and information technology (what we see as potentiality) that, brought together and put in the context of energy, can and will usher in a new era for us. Because now we can have energy that is not only no longer material, it is also no longer simply abstract, it is now *symbolic.* Let us be precise about this: the energy itself is still exactly what it was, it is still as physical as it has always been. But the way we handle it, what you might call the conditions and constraints around it, our management, so to speak, of it, has now turned symbolic: we can endow it with *meaning.*

As long as energy is heat, motion, light or matter, it is only these things, nothing more and nothing less. But when we convert it into electricity, energy can behave like data, it becomes *medial.* And if it is possible to handle energy not at the material, physical level, but at the symbolic, medial level, then we find ourselves in a different dimension.

The limitations we so tangibly come up against in the context of energy are all of a *material, physical* kind. Yet where there are no pre-defined, no absolute limits, none at least that we can as yet identify, is in the realm of *virtuality.*

You can tell that now things are starting to get complicated. On the one hand. On the other hand they are starting to get wonderfully simple. But we will want to get our heads around it all, and so let's take one step at a time and proceed.

II   A PATH OF RATIONAL OPTIMISM

This is not a project. It's certainly not 'our project'. Much rather, we are observing what's happening, and the question for us isn't whether it is happening or not, because clearly it is, but *how* does it happen, and how do we relate to this development, what is our role in it; how do we shape our world, what do we create in it. This is one of the great, and greatly empowering characteristics of the era we've entered: that it neither has to, nor wants to, be steered from central controlling bodies, such as governments or government appointed agencies. It can grow *organically*. As more and more people adopt a technology, it becomes cheaper and more versatile, more capable all the time. As more and more people connect to a network, they invent what it does, they make it what they want it to be. The investments that need to be made by service providers, energy companies, and suppliers to the network are covered on a 'need-to-have' basis by the consumer, at prices the market sets in accordance with, by and large, supply and demand.

This means that patterns similar to those in other electronic consumer technologies will be observed: early adopters who recognise the benefits of the technology and who are, perhaps for idealistic reasons or out of experimental curiosity, prepared to pioneer it, buy into it first and at a higher price, setting the territory for the mainstream market to follow. As soon as critical mass is reached, prices rapidly decline and the technology proliferates.

This does not mean that some important policy decisions do not have to be made, or that major, large-scale investments are not necessary. Clearly they are. In order for a genuine global energy network to come about, we will need updated electricity grids that fluently transfer energy through high voltage cables across large distances, maybe continents. This is not cheap, but it's possible.

Mobile telephony once again serves as a workable example for how a largely consumer-driven rollout of technology can even the path for infrastructural changes on a gigantic scale. The fact that today every village, every town and every corner of every city across the majority of five continents receives a mobile phone signal is attributable entirely to the fact that people want it. And because they want it they are prepared to pay for it. Governments, rather than putting out subsidy programmes for preferred industry sectors, sold off licences to credible bidders. The public purse, instead of being depleted, was being bolstered by commercial enterprise that actually benefits the user. And the user, through subscription and phone charges, foots the bill.

No approach is without its problems, and this certainly applies here too: for all its open market characteristics, it was in fact European regulators who had to tell mobile operators that they must stop ripping off their customers when they travel across national boundaries, for example. Big companies left entirely to their own devices are apt to eventually stop being innovative and dynamic, and start being greedy. And energy is no less important a general provision than communication. So there will have to be safeguards and regulation. But that is not the same as central control.

What we are trying to show is that where there is a demand, changes – even large, conceptual shifts – can happen very quickly, and they can be user driven and financed. When it comes to energy, it helps us to bear this in mind. Our position has for too long felt like that of the rabbit, caught in the proverbial headlights. But being run over is far from the only possible outcome, in our position or his. There is a massive shift underway now, and the public sector as well as private companies are beginning to embrace it.

Let's park, for a moment, the thought of energy abundance: we agree that this is not going to come about overnight. But what is now underway and what we can, by our own brain and will power, accelerate and take advantage of, is this transition from one technological generation to another. It benefits us greatly to be aware of this and seize the opportunity. We are not being flippant nor are we being naive when we ask for the tenor of catastrophe to be replaced by one of opportunity. We are not being unrealistic nor are we misguided when we say that now is the time to rewrite the equation and reposition resources within this equation, so that they no longer serve us for our energy needs and are therefore being used up, but can instead be used for our material needs, and endlessly recycled. We are not being smug nor are we being hypocritical when we say 'leave ideology out of it'. And we are not being idealistic when we say that there is a new type of autonomy and emancipation to be found in a networked world.

What is most useful in a period of great change is creative thinking, flexibility. The willingness and agility to try new models, different approaches. Obviously, this carries the risk of mistakes and failure; there will be bumps in the road, that is to be expected and it would be foolish to think otherwise. But a challenge is not the same as an insurmountable obstacle, and these risks are worth taking, compared to the risk of standing still and being flattened by our own ineptitude.

So the path we're proposing is one of rational optimism. You could also call it *applied* optimism. It means not using optimism as justification for doing nothing ('things will turn out fine'), but fuse an optimist attitude with both the spirit and the practical application of invention ('yes, we can do this').

We propose a path that uses and embraces the potential of digital technology and that makes use of the extraordinary, vastly surplus-to-requirement energy we so very obviously have at our disposal. And this, it is really worth stressing, is a way forward, not back. It's a technological path, not a luddite one.

The technology, meanwhile, is not exclusive. In this, it differs again very much from some of the other technologies available. Nuclear power, hydrodams, 'clean' coal, tidal power – for all their various advantages and disadvantages, they have one thing in common: they are expensive, heavy machinery dependent, high maintenance, labour intensive.

One may argue about the level to which any of them is part of a 'solution' as opposed to part of a 'problem', but that is by the by. What is certain is that they all require a lot of hardware, a lot of expertise, massive health and environmental safeguards and a whole lot of money to get going.

With photovoltaics, you need a piece of PV foil, a couple of wires, maybe a battery, and you're online. Total cost, including the laptop, dongle and pay-as-you go access to the internet: €500. You can work, you can communicate, you can start an enterprise, you can employ people. You can study, you can design; you can sell, you can buy. You are right there on the current crest of the social and technological wave. And you can do this and be there whether you are in Shanghai, Delhi, Mogadishu or, as we have seen in our case study, Addis Ababa.

So this certainly is not 'no-tech', but it's low-tech enough to get you started. Then, gradually, you may want to make things more sophisticated. As we have also seen, there is some distance to cover between the basic set-up that gets you online and the fully-fledged 'intelligent household' that connects your whole existence to the network. But the principle is the same.

What may be holding us back most at the moment is that we're inclined to look at an entirely new, wholly different thing and apply to it old thinking. And so we get stuck up on red herrings. We ask questions such as: what if the sun doesn't shine? What about storage? What about the raw materials for solar foils? What about the energy it takes to make them? These *sound* like reasonable questions, and in the old paradigm, if you forgive the expression one last time, they would be. On what we have referred to as the 'balanced planet', they are of paramount importance. But on the plateau we are now getting onto, which we are calling the 'genius planet', they become peripheral. Because in a connected energy network, storage is not the big issue. Yes, there is the need for some short term storage, and rapid progress is being made in this field too,[CLXXXI] but just because earth has stored so much energy for so long for us does not mean that we have to continue to do so now. Growing, pan-regional networks with a radius of up to about 2,000 kilometres (roughly 1,200 miles), localised domestic heat pumps or batteries, supplemented by the buffer provided by electricity storing devices, such as electric cars, of which there will be millions, existing power generating techniques, and new solutions such as solar fuels will combine to address the storage issue to the extent that it is one.

Also, remember that the transition period that we envisage is one of about 25 years or so, and it will encompass a number of solutions that may not necessarily be part of the long term picture, but that will, for the duration, be able to provide adequate cover. An example might be a number of small gas powered 'top-up' stations that can kick in at short notice and supply a burst of energy to the grid when required. This would neither present a significant environmental hazard nor would it constitute a massive rear guard investment. The technology for this kind of safeguard is fully understood and available and the infrastructure for it is in place. We've mentioned solar fuels: the potential these offer for bridging supply gaps and boosting energy supply in the short term (for example in small solar gas power stations) is as yet wholly unexplored.

So there will be back-up options. But they are comparatively minor in scale, and as the 'smart' and 'clean' technology settles, improves and strengthens, there will be less and less need for them. We can, once again, refer to information technology as an analogy. The transition from handling and storing information in material form on index cards, in reference libraries and on microfilm, to handling and storing information in 'the cloud' took approximately one generation. During that transitional period, neither paper, nor magnetic data storage, nor external hard drives or built-in hard drives disappeared overnight. But as the transition took place, they became less significant. Within a few years' time – for some of you who read this already today – the data storage capacity of your laptop will be of secondary or tertiary interest. What will be of primary interest will be its design, its interface, its processing speed and very importantly the reliability and speed of its possible connections to the internet, specifically the 'cloud' of servers where the vast majority of your data will reside.

And here, too, we don't do away with hardware and infrastructure: the servers and the high-speed internet connections that allow you to carry around with you a lightweight tablet or laptop computer with almost no data storage still all have to exist. They have to be wired up and powered. They have to be manufactured to highly reliable standards, and maintained. But there is really no limit in sight, for the foreseeable future, to the capacity of information that can be handled in this way. And that is the crucial factor. With material-based data storage, you get to the limit of what you can usefully, operationally do: there comes a point at which a reference library based on index cards is so unwieldy and slow that it no longer can effectively fulfil its purpose. There is no such horizon yet for virtual data. Nor is there any such horizon for 'virtual' energy.

Similarly, raw materials for solar foils: quite apart from the fact that the most prevalent ingredient is essentially sand, of which there is so much in the world that we'd need to shift deserts to exhaust it, advances are being made, as we have seen, into other ways of converting solar power into electricity. This is still relatively new and we're as yet early in the process of developing the technology, but we need not get caught up in arguments about what happens if and when there is a shortage of solar foils: there is no foreseeable end to either solar power reaching us nor to us having the means to use it. And *what about* the energy it takes to make a solar cell? In the longer term, it doesn't really matter. As long as during its lifetime it garners even marginally more energy than it took to make it, it's a winner. We don't have to assume that we will forever be burning fossils to make solar panels. That's a not very fortunate transitory situation. But once the majority of energy comes from infinite sources, making another bit of kit that helps getting at some more of these infinite sources can only, by definition, increase the overall availability of energy.

III   "THIS IS OF THE DEVIL" – FACING OPPOSITION

If everything we talk about here is technologically possible, and economically viable, then what's the obstacle? Who is against it? And why?

As it happens, there are any number of people and groups who are either actively against the path we are proposing, or whose doubts about it are so pronounced that they come to effectively stand it its way.

As we've briefly mentioned, while we were writing this book, Ludger and Vera were travelling extensively around Europe, North and South America and Asia, talking to people, giving lectures and hosting seminars, learning a lot themselves in the process, and finding a great deal of support for their ideas, which are also being developed, taken up and exchanged at many other institutions around the world. But there have also been some fairly extreme reactions. At one such event, a young CEO of a 'green' energy company with a turnover of €6 bn got up on his feet and declared: 'If what you say here is true, then this has to be banned. This is of the devil.' It was an unusually vehement response, but it served to remind us that where we see a development that's underway and that is to be embraced and welcomed and worked with, there are people who very much see a threat. And some of them are resident in surprising quarters.

We would not find it particularly astonishing to realise that someone whose existence largely depends on, and who has maybe spent many years or decades working in, the traditional energy industries should be inherently sceptical. What's more perplexing is that on the environmentalist side of the debate, among people who take an active stance against the type of industrial pollution that stems from fossil fuel power stations, you quite frequently encounter a resistance to what we might term 'technological progress'. (The word 'progress' doesn't sit so easily, because it is in itself weighted and fraught with difficulties, but for the sake of everyday understanding we can certainly call what we are putting on the table a technology-driven approach aimed at moving us, as a society of societies – as a species, if you will – on.)

While it may not be representative of all shades of the political spectrum that labels itself 'green', there is a fairly wide-spread assumption among the Green caucus that what we need to do is *restrict* ourselves. Slam on the brakes and decelerate. The idea is to go *back to nature,* in a sense, and do things more simply and in an 'eco-friendly' manner. Quite often, when we talk about there being room for more growth, of there indeed being a *need* for growth, we get horrified reactions from very well-informed people who see this as the worst possible suggestion we could make. We have already argued our case for growth, for expansion and for assisting developing countries to become fully networked, technologically up-to-date societies; there is no need for us to argue it all over again. But it is worth bearing in mind that this issue divides opinion, and not necessarily along obvious lines.

A lot of people advocate new technologies and carbon neutral methods of energy generation and focus very much on the physical, material aspects of resources on the one hand, and on emissions on the other. With this, they calculate and are able to scientifically prove, as Ernst Ulrich von Weizsäcker [CLXXXII] does, for example, that we can achieve a 'Factor Four', and explain how we can extract four times the amount of wealth from the resources that we currently use. Similarly, Al Gore with *Our Choice,* his follow-up to *An Inconvenient Truth,* is able to demonstrate how, using a combination of 'alternative' and 'renewable' energy sources, we can wean ourselves off fossils, drastically reduce our carbon output and still keep economic development going. [CLXXXIII]

The Global Marshall Plan, also initiated by Al Gore, meanwhile, seeks to achieve a "World in Balance" within a framework set by the United Nations Millennium Development Goals. [CLXXXIX]

These are informed calculations and well-intentioned initiatives, which we don't mean to dismiss sleight of hand. But not only are they conceptually unadventurous, they also don't go — we believe — far enough, fast enough. Because we don't want a world in which the developed, post-industrial countries can continue to grow by some, and some of the emerging economies can grow by some too, but large portions of the global population are effectively kept where they are now. That's what a 'Factor Four' entails. Even the "World in Balance" sought by The Global Marshall Plan, whose adopted stated Goal No 1 is to "eradicate extreme poverty and hunger" will take a very long time indeed to come about with its benevolent but conservative approach. The World Bank estimates that in 2013 (the latest available set of figures) "10.7 % of the world's population lived on less than USD 1.90 a day," which was down 35 % from 1990, but still left 767 million people in what it defines as 'extreme poverty'. [CLXXXV] Poverty, it sounds obvious, comes from economic disadvantage. From not having enough of a share of the global economy. From not having energy. If you want to lift people out of poverty, you need to give them the means to cultivate their land, and once they are able to feed themselves they need the means to cultivate their cities. Energy.

Hsilo, Taiwan; Tan-Tan, Morocco; Annecy, France; San Joaquín, Venezuela; Redcliffe, Australia; Cumbernauld, Scotland; Faro, Portugal... What do these places have in common? They're all towns of roughly 50,000 people. Imagine a week in which you turn on your television every day, and every day you hear of a town wiped out: on Monday, Hsilo; on Tuesday, Tan-Tan; on Wednesday, Annecy; every day a whole town somewhere in the world, killed off through manmade catastrophe. If it were terror attacks, you would be *outraged*. If it were poison in the water, you'd demand for heads to roll. If it were war, you'd call for action. This is what's happening: poverty, today, kills roughly fifty thousand people a day. It's a difficult figure to be sure of: how can you be certain that someone's death is caused by poverty, directly or indirectly? What you can be sure of though is that it is not a natural disaster. Poverty, by now, is due to bad planning, bad policy and inequality. We can end poverty. But only if we provide energy and allow for growth and development.[CLXXXVI]

And that's why we need a different type of thinking. Of course, any idea that contributes anything constructive is welcome, but we do not need thinking that in essence leaves everything as it is but makes our resources stretch a bit further, we need thinking that frees us from resources and gets us onto a new level. What we need is a *fundamental re-conception* of energy. The problem with so many of the proposals and plans on offer is not that they are bad, the problem is that they are not good enough. Because even if, instead of coal or oil, they use biofuels, or wind or water, they think of energy in the same qualitative terms that we have been using for centuries: they 'produce' the energy, by means of mechanics, then distribute the energy. And one of the methods of 'distribution' is electricity. But remember, electricity is not just a 'method of distribution', electricity is an *abstraction* of energy. In a moment, we'll be talking a bit more about symbolisation and why it is of relevance to the energy story. For the moment though suffice it to say that what we envisage is treating energy as something *abstract* from the start. And for 'abstract' you may read, in this context, 'digital'.

We are once again over-simplifying a bit, but if you compare the move that's in progress now to the move from analogue to digital in music or imaging technology, you will get an idea of what we mean. What changed the game there completely was not just a new, more sophisticated, higher quality recording and processing technology that became more affordable as more people bought it. What changed the game completely was the move to 'digital'. From sound or image capture, via processing to output and distribution, everything today happens at a digital level, which is an *abstract* level, and that's what has given us such overflowing wealth in content. Apply the same principle to energy and you get a feel for the potential we have at our fingertips: if you start thinking of energy in exclusively abstract terms – cutting out mechanical conversions, physical storage, 'analogue' processing – you no longer have to think in terms of Factor Four, or Factor Ten, for that matter. You can forget factors altogether and reposition your horizon somewhere you've never even looked before. And that's what we're after: a real change in perspective.

So probably the people most likely to stop us from taking this path and successfully arriving at our proposed destination (which is, of course, like any destination, always only an intermediate one: there will be appearing, on our horizon, long before we reach our next 'goal', a new set of potential destinations), is us. Because although everything described in this book is based on existing, working technology and sound theory, following it up actually requires almost an act of faith. Faith not in a god, the universe, or in the forces of nature, but faith in ourselves and our abilities. And 'faith' is not a very scientific term. It doesn't, in justifiably hard-nosed business circles and among academics professionally given to enquiry, inspire a lot of confidence. But confidence is exactly what we need to have, if we want to make this happen. Because the alternative is permanent segregation: an apartheid of haves and have-nots, where there are those who have energy, technology, science, education, health care, and those who don't.

IV   THE DIFFICULTY OF THE NEW

Invent a super new thing and put it in front of 'the people'. What will happen, almost immediately, is that 'the people' will divide into broadly three distinct groups: there will be those who *love* your new thing, if nothing else because it's new. Some will want it straight away, and will be prepared to pay a high price for it, and talk to their friends about it and rave about it online, and even if the new thing is far from perfect, they will overlook its imperfections because these pale, in their enthused sight, into insignificance by comparison. In technology marketing, these are the ever crucial 'early adopters' who willingly and consciously help finance a lot of technology development, because they are happy to pay over the odds for getting new things first.

Next to them you will have the indifferent. They have seen new things arrive on the scene before and they are not convinced that 'new' equals 'good', let alone 'better'. They take a 'wait and see' approach and think to themselves, if everybody still talks about this marvellous new thing in two or three years' time, when my old thing may perhaps come up for replacement anyway, then we'll have a look into it, why not.

And then you have the ones who will *loathe* your new thing, principally because it's new. 'Not another *new thing!*' they will cry out, 'why can't they just leave us alone with their perpetual inventions that never really change anything anyway but just make life more *complicated*.'

There is, of course, a whole science dedicated to the diffusion of innovations. In a nutshell, the adoption pattern of any new thing is that of a normal bell curve, with a tiny percentage doing the innovating, which is picked up by the early adopters, perhaps ten to twelve percent of the population, who are then followed by the 'early' and 'late' majority, who together make up well over two thirds, with the 'laggards', about another 15 % or so, appropriately behind, and the rest coming on board either very late, if they have to, or indeed never.[CLXXXVII]

What is true for new things is also true, almost exactly, for new ideas. Which is why almost any new idea will almost certainly and almost immediately have its evangelists and its detractors and, in-between, those who sit most comfortably on the fence. Whether the idea is one of spirituality or one of science makes almost no difference. And there is inherent, in anything that's new, the question 'how real is its novelty?' – is what's being put on the table really new, or is something old merely newly packaged and presented as new. And this, again, is largely down to interpretation and the kind of stance you take towards things generally, new and old.

When, on 5th October 2011, Steve Jobs[CLXXXVIII] died, something truly fascinating happened in the universe of social networking and micro-blogging. Some, indeed many, mourned the passing of one of the great visionaries of the turn of the century and eulogised his achievements as the *invention* of creative and user-friendly computing. This was watched, read and commented on with some bemusement by a constituency who had never 'got' what Apple was about and couldn't relate to any of what was being said. For them, Steve Jobs was a man who made some nice-looking gadgets and knew how to market them.

Considering just how radically different they are supposed to be, there is a remarkably fine line between science and religion. The proclamation of a Kingdom of Heaven on earth through the arrival of the Messiah is not, in essence, all that different from the proclamation of peace and well-being for all mankind by virtue of a groundbreaking technology. Our own excitement about the latter notwithstanding, the reality is always the same: anything that's new, be it a product, a design, an idea or a way of thinking, has to be continually assessed, reassessed, positioned, repositioned, questioned, considered, reconsidered, rephrased, analysed, reviewed, evaluated. And as we do this, we realise that there is no Kingdom of Heaven on earth, nor is there instant peace and well-being for all mankind. What there is, is a slow continuation, development and formulation of our culture.

Perhaps it is more sensible then, rather than celebrating the new and proclaiming it, to celebrate our culture, and proclaim that. Because what brings us the new, and what at the same time enables us to make sense of the new and test it for its validity and adopt of it what is of value and discard of it that which isn't, is our ability to use critical faculties and pitch them against our thrill of having something to get excited about.

All progress really is, is our capacity to learn. And this would suggest that we need neither be fearful of the new, nor do we need to exult it, as if it were able to spell the end to all human suffering. What we do need to do is *talk* about it. Articulate it. Listen to how other people articulate it and discuss it. Because that's how we learn to differentiate, that's how we learn to appreciate and use to our advantage complexity, instead of dismissing it as a mere inconvenience and looking instead for simplicity where it doesn't exist. It means finding our own voice and letting it be heard in this ongoing, global conversation that is taking place, and in doing so creating, defining, nurturing values.

This is what you could call 'serious storytelling' – putting ourselves into and making sense of a meaningful narrative of our own history and culture.

V   LEAPS OF THE IMAGINATION

At one point, about twelve thousand years ago, our forebears embarked on an incredible project: they decided to leave the comparative 'safety' of their caves in the forest and stepped out into the open landscape. Here, they were far more exposed and apparently far less in control, because suddenly above them was the weather and all around them were potential predators and dangers. And here, they decided to settle: to cut down some of the trees and plant types of grass instead that they could harvest and turn into bread and eat. And other types of grass that they could feed to their livestock. They decided to keep livestock, instead of hunting beasts, and to build houses, not just for themselves but for their animals too.

It was a bold move they undertook, our ancient ancestors, and it had the most astonishing results. Far from being drowned by the rain, parched by the sun, blown away by the wind, and eaten by the wolves, they thrived. Around about 10,000 BCE, there were somewhere between about 1 and 4 million people on earth. By October 2016, there were a bit over 7 billion, and by about 2060 the world population will have pipped the 10 billion mark: it will have grown to ten thousand times what it was when we started to settle.[CLXXXIX] And our existence today bears no comparison to that of our early pioneers. We live, in every possible respect, better, safer, more comfortable, healthier and much longer lives than they did. The adventure they undertook has truly paid off, and spectacularly so.

Where exactly would we be without leaps of the imagination? Where, without curiosity? Without daring to leave the known behind and step boldly into the new? Would going to the moon have seemed too far? Would we have stopped short of taking to the air? Or would the idea alone of staying on the ground but travelling faster than the wind have seemed so improbable that we simply wouldn't have tried. Maybe we would have felt it inconceivable that there is a way of talking to each other through a wire, or, over many thousands of miles, *without* a wire.

Of course, the daring of invention is incremental: one builds on the one that has gone before. But occasionally it's easy to forget that it's as old as the human experience itself. To prehistoric man the idea of settling would have seemed fantastical. Grow stuff? What on earth for, there's stuff all around. Settle? How? You'll run out of things to eat in no time, you need to follow the food. Build houses? What good could possibly ever come of that? It would have seemed idiotic. No, better, surely, to stick with the good old tried and tested ways of doing things: pick up a club in the morning and chase the mammoth till you've caught it, then kill it and eat it. That was a good day's work and plenty to be getting on with. No need to start thinking in terms of newfangled dwellings and what have you. And who would eat grass anyway, grass is for rabbits...

Back then, twelve thousand years ago, the average life expectancy was about 20 years. By the mid-1960s, around the time IBM launched the first computer family, System/360, which arguably marked the beginning of computing on a broad scale in a meaningful way, the average life expectancy, across the globe, was well into the sixties.

What would life on earth be like if we had a lot more energy than we need? Not just twice or three times as much as we use at the moment, all of us together, but hundreds, thousands of times as much. And what would it mean if this vast amount of energy were not ever going to run out, but just came through, continuously, all the time, with absolute reliability, free of charge from the supplier, for as long as there is life on the planet?

We would not think of ourselves as having an energy problem *as such*. If someone said we must reduce the temperature on our thermostats by a degree or two and not make unnecessary trips into town, we would take them to one side and say to them: 'Please be serious.' The very idea of turning off all the lights in every city on earth once a year for an hour or so to make a point about 'saving' energy would appear to us preposterous: what on earth would you save it for? The energy would still be there tomorrow, it would be pointless to save some of it now so you'd have more of too much the next day, and besides there's nowhere really to put it, because it's not a tangible thing, energy, it's not a substance, it's energy. It wouldn't cross our minds to say, 'let's save some energy and not drive into town today', that would be like saying 'let's save some air and not go for a walk in the park today.' It would be ridiculous.

Change happens all the time, whether we are consciously aware of it or not, often whether we like it or not. And much of the time it is much easier to see the significance of change with a bit of distance, after it's happened, than while it is actually in progress.

When we started working on this book, at the beginning of 2009, the idea of smart grids and smart meters, of generating some energy on your rooftop with a few PV solar cells and feeding it back into the grid if and when you didn't need it, was being touted around and discussed, but it looked and sounded like something that would happen some time in the mid-distant future, one of those things that may well come about one day, but that one could probably ignore for quite a while yet. Within three years, every major energy provider in the UK was offering precisely this. It is now reality. Depending on where you are in the world, and when you read this, you may already be getting regular promotion material for something that we, only a short while ago, thought about as a concept; in fact, you may already have subscribed to it. The world moves fast when it wants to. We do well to move with it.

VI   POWER STORIES

Let's take a step outside the narrow confines of our subject. Let's do what in business circles they like to call 'thinking outside the box' for a while. Or, as Monty Python might put it, go for 'something completely different.' Except, of course, it isn't completely different, it has very much to do with what we've been talking about, but we want to talk about it in a slightly different way.

Let's take up this idea of a narrative, of 'serious storytelling'. Why? Because isn't that what we do, all the time? Isn't what we're doing, day by day, hour by hour, minute by minute, invent our own story? And we use the word 'invent' here deliberately, because we are, quite literally, making it up as we go along. All of it. Our entire existence is nothing but one great big invention. From Stonehenge to Twitter. We come together, we find ways of expressing ourselves, we live and we die. Stories is what we're made of. Without them, we are just an organism that may or may as well not exist. And some will argue that that's exactly what we are, and that once we are gone from this planet there will be nothing and no-one to miss us or to even know that we've ever been here. That may or may not be so. But while we are here, it is stories, all our stories, be they 'fiction' or be they 'fact' – and where, indeed, do we draw the line between them? – that make us who we are.

And if that's the case, then why not tell stories? Why not make them up? With a new flourish, with a new sense of purpose. Certainly with a new imagination. Who's to say – whoever did say? – that science and fiction don't mix. That dreaming and building a reality are incompatible with each other. That fantasy and engineering don't go together. Who's to say that history is one thing and invention is another. Isn't *everything* an arrangement? The putting together of things – facts, ideas, details, experiences, feelings, thoughts, emotions, perceptions, opinions, dates, outcomes, projections – in ways that make sense?

We've made a point of trying, as much as we felt we could without seeming rash, to avoid statistics because statistics can prove or disprove virtually anything, and we've also mostly avoided scenarios, largely because more often than not they turn out to be wrong. But when we talk about 'serious storytelling' we do mean working with tangible possibilities.

Take farming for example. Farming stands right at the beginning of what we call our civilisation. It's cultivating the land, working with nature, taming it to some extent, but learning from it also. In our Case Studies we've illustrated how farmers in Switzerland could augment their income and create whole new economic realities by getting involved with energy 'generation'. In Bavaria, Germany, this is well underway: dairy farmers have cottoned on to the fact that their land yields up more than just feed for their cows, and so they started to get into multi-purpose farming, using some of their land and stables' rooftops to collect photovoltaic solar power, dotting wind mills around the pasture and burning bio gas to produce electricity. We've elsewhere in this book not been particularly enthused about bio gas, and we know of the reservations there are about wind mills, but what's interesting here is that a new story emerges: the story of farmers who are not just cultivating the land and using nature as a secondary or tertiary source of energy, but cultivating energy itself. They become dairy-and-energy farmers. And what else is food – be it dairy produce, meat or wheat – other than fuel for people. Maybe farmers have always been energy farmers. Maybe all farming is, is getting at the energy from the sun, in one way or another. Who's to say that that's bad? Certainly, some of the big electricity companies in Bavaria have been reducing their own output, because of the amount of energy they're getting fed into the grid from their farmers.[CXC]

Compare that, then, with somewhere like Saudi Arabia, for example. We associate it with oil and desert, yet even optimistic forecasts by the country's own government expect the oil to run out by the end of the century.[CXCI] And whether or not that will be the case, who's to say that where there is desert now there can only ever be desert in the future? Clearly you can use the desert to produce solar power, that's an obvious and simple solution, and it shouldn't surprise us if before long Saudi Arabia were to follow its small neighbour's lead and, like the United Arab Emirates, diversify into solar. But by the very same token that permits us to ask why dairy farmers in Bavaria should not also be solar farmers, we are entitled to pose the question: why shouldn't solar farmers in Saudi Arabia become dairy farmers? Is greening the desert so unrealistic? Saudi Arabia has access to thousands of miles of coastline on two sides of its peninsula, and plenty of sunshine. By instigating a slow but gradual solar-powered desalination programme, would it not be possible to irrigate the land from the coast inwards? Or what would happen if it were possible to create oases? Can we picture it? We can certainly imagine it. All that's required is that we allow ourselves to rewrite the equation. In the case of the Bavarian farmers: bring energy into it: see what new results come about. In the case of Saudi Arabia, bring water into the equation: see what happens. And this is something we can model: a solar panel little bigger than a grown man's palm (10 cm × 10 cm, or a bit more than 9 square inches) is capable of desalinating 1 litre of sea water per day. This means that one average 'solar tree' is able to provide you with the water you need to keep between twenty and thirty natural trees amply watered all year round at roughly the same level of precipitation as is available to the average tree in Bavaria: 1,400 millimetres annually. This does put things rather into perspective: are we really going to fight wars over water, and continue to behave as if the only possible outcome of man-made technology was extreme weather, global warming and desertification when all it takes to make a small forest grow in the desert is some solar panels, some micro filters, some cabling, a pump and some freshwater plumbing?

The point is of course always the same: it doesn't matter so much what the specifics are of what we want to achieve, what matters is how we go about it *conceptually*. As we've said before, we like to think of it in terms of applications. Why applications, again? Because that's what you use in computing, in information technology, on your smartphone. You say to yourself: if it were possible to do anything, what would I do? Or you see something and you say to yourself, wouldn't it be amazing if I could do this? And you create an application for it. You assume that the platform is ready for your idea, and it is.

As long as we think in terms of resources, we subject ourselves to scarcities, to the laws of precedent and to restrictions. 'This is what we can do with so much oil.' The moment we start thinking in terms of *applications* we can begin to actually cultivate new potentials. Because we can think in ways we've never thought before. We have no idea what's going to be possible. Yet. But we have no doubt, we'll find out. It's exactly what we did in telephony, it's exactly what we did in computing.

Which brings us on to a subject that we've flagged up once or twice now, but haven't really got into yet. But now is perhaps not a bad time to do so. Because we've drawn this link, this connection between energy and information technology, and tied it up fairly tightly. We're beginning to be able to imagine the house as an application inside its thermal field. Rather than in isolation, we are considering ourselves *connected*. And being connected, we find ourselves at once surrendering some of our control, and at the same time gaining a whole lot more power. It seems paradoxical that that should be so: how can you become *more* powerful when you surrender control? But that is exactly what we are doing, just look at YouTube, or Wikipedia and you can see what we mean by gaining power by surrendering control.

So now we want to go one additional step further. And that is into *symbolisation*.

VII   THINKING ENERGY

Having dealt with most of the practical and technical aspects that we want to talk about in this book, let us take you then on a little trip into the realm of theory. We won't get lost in it and we won't leave you stranded, but given this opportunity we have of sharing with you our thoughts, it would be a pity not to touch on at least the surface of the actual thinking that lies behind all this.

We postulate that the boundaries, the limitations, which we are reaching are not primarily of a material kind, even if we feel the consequences of reaching them on a material and indeed existential level, as we are aware we do. What is that supposed to mean?

The earth, with all its plants, with the animals and ourselves, with the weather, and all the energy that we as humans take out of the ground in the form of fossil fuels, or grow or harness from wind, water and the sun directly: we look at all of this as a system. Strictly speaking and in a philosophical sense, it isn't really a system, but most people would regard it as one and would also agree that it is an intricate, complex set of interdependent factors that forever mutually affect each other. So whether it is a system or not, what we are certain of is that this planet we live on, with everything that's on, in and around it, can appear at once extremely fragile and awe-inspiringly robust.

A widely held, and perfectly understandable, assessment of our current problems holds that this set of interconnected factors has reached the boundaries of what it can sustain, that we are at breaking point. We have to curtail our consumer behaviour, we have to save energy, we have to halt the increase in population, we have to end growth. It is not an arbitrary view, and it is certainly not irrational. And it is very much rooted in precisely the fact that we tend to view the world as a 'system'. From it stems, with some sensible logic, the conclusion that we have to 'balance the system'. It is the pervasive and compelling view of the 'balanced planet', as we have seen. In the famous 1972 study *The Limits to Growth,* and its 30-year update in 2004, this point is made, and calculated through in great detail.[CXCII] And it amounts to just this: there is, as the title suggests, a limit.

The concept of 'sustainability', and indeed our use of the word, ties into this directly, and again for very good, sensible reasons: there are material, tangible limits to how far we can go; the science says so, and so say the statistics, and we very much get the sense that those limits have now been reached, that we're right at them. The oil is running out, the coal that we're burning is overloading the atmosphere with CO_2, water is getting scarce, food is getting scarce, space is getting scarce, we need to stop. And in our day-to-day existence many of us – more and more of us, it would seem – are experiencing these material, tangible limitations, or at the very least we appear to be experiencing them: energy prices are soaring, summer temperatures are rising, glaciers are receding, polar bears are floating lonely on broken-up ice shelves, wherever you turn you are told that you must reduce your carbon footprint. It does indeed seem to us that we've reached the boundaries. We can see them, feel them, we are asked to pay for the measures that are being taken so we don't step over them. It is *real.* Or is it?

Is it possible that the reason we have reached our limitations is not because they are actually there, but rather because of the way in which we look at the world and the way in which we understand it? Is it possible that we could find ways by which these boundaries, that seem so very real and insurmountable, are in fact nothing more than the boundaries of our imagination and our capacity for inventing new concepts? And how could we find out? How, indeed, could we, if we found out, make use of that discovery and overcome these boundaries completely; not put them a little further out, so that the oil would last a little longer, for example, or the sea would rise a little slower or the food would feed a few more people, but so that they become immaterial, these boundaries, so that we can ignore them, so that we can expand and grow and develop and evolve and create until we hit the next set of boundaries, which will be of a totally different kind that we can't even think of yet?

VIII   OUR EVOLVING TAKE ON ENERGY

If we go back for a moment to our trusty log of wood that has served us so well over the millennia: what is it, this log, exactly, and how do we relate to it? In the beginnings, when we first started to use fire, the log was there in the forest and we picked it up. We just used what we could find. Energy was nature and nature was energy. And for a while, it sufficed. Then we started to settle, and the log of wood was no longer something we just found and picked up, it was something that we *cut down.* We cultivated our use of nature by applying instruments, tools, to make it our own, and very soon we began to realise that a good way of making sure there'll still be a log to cut down tomorrow and the day after is by *planting* trees: we cultivated our energy 'generation' too. Except that planting a tree is not, as we've seen, 'generating' energy, planting a tree is 'collecting' or 'harvesting' energy that's there anyway: the energy that comes from the sun. But still, the reason we keep running out of energy is that we keep using up the log of wood: we're using material, physical stuff to get back at the energy that comes to us not in the form of material or physical stuff, but in the form of heat and light.

We obviously didn't stop there. And the reason we didn't stop there is because the uses we had for fire, our demand for it, rapidly out-grew the capacity of wood to provide it. So we found other things to burn. Coal, oil, gas. But this was not so simple as we now make it sound: coal, oil and gas have always been there, but it took a considerable amount of ingenuity (and more than a little trial and error, for which many paid a very high price) to get at them, make them safe and us-able. This step too is of great significance, because we no longer simply took stuff we found lying around or were able to grow and chop down; we now manipulated and processed, refined and purified stuffs so we could get at the energy in them. There is a conceptual leap in that, too, and one which unleashed tremendous amounts of energy. Although coal had been lying around for millions of years, it was only really during the industrial revolution when we learnt to handle it properly that we were able to extract from it around a hundred times as much energy, per like for like quantity, as we'd got out of wood. But coal, oil and gas are still nothing more and nothing less than solar energy, stored in stuff.

With photovoltaic solar power, and interestingly with nuclear power too, we get away from burning stuff and instead start to use the principles of quantum physics, albeit in very different ways. Photovoltaics allow us to use solar power directly, without making the millennia-old detour via fuel, while nuclear power allows us to use possibly the only type of energy source we have, apart from geothermal and tidal, that does really not come, in one way or another, from the sun. It is the most fundamentally earth-bound type of energy there is.

What this means is that, while we are not much closer to being able to imagine or express what 'energy' really is, we have in fact moved on a fair bit. We have long since started to *cultivate* our use of energy, and we have, though only over the last century, begun to understand the behaviour of energy at atomic and subatomic level. And this has come in tandem with – and has given us access to – possibly the most brilliant thing we've ever got to grips with: electricity. Because it is electricity that allows us to instantly distribute energy into any corner of the world, it is electricity that allows us to put the energy that we have to not just a few dozen but to a few million uses, and very crucially it is electricity that allows us to *digitise* information, and it is electricity that allows us to *symbolise* energy.

A symbol is an abstraction of a thing, and so that's exactly what we now want to discuss in a little more detail: the idea of electricity as an *abstraction* of energy.

IX   ABSTRACT ENERGY

It does not sit so easily, this notion of electricity as an 'abstraction' of energy. What, you are entitled to ask, is an 'abstraction' of energy? Is energy not either energy or no energy? How can you make energy abstract? And perhaps here we should qualify. Maybe 'abstraction' is not so helpful a term. Because do we mean that energy becomes an idea of itself? That it can no longer do what it could do before it was 'abstract'? No. What we mean is that it has become removed from its source and can now be dealt with in a non-material way. It has, effectively, become a meta-form of itself. *Really?* Well yes.

Take the log of wood. It's a physical object, which we can stack up, chop in two; touch and smell. When we burn it, we know it's burning, there is no doubt about it, it's visible, tangible. In order to get it from the shed to the fire place, we have to go outside and lift it up and carry it in. If we want to make a big fire, big enough, say, to move a locomotive around a track, we don't use wood, we use coal, which is fossilised, very old wood, and we burn that instead. Instead of going to the shed we go to the coal mine and dig it out from there. To get it from the coal mine to the train depot we load it onto lorries and trains.

How does this compare to an electric fire, or an electric train? What can you actually see, or touch here? Nothing. The electric 'fire' is no fire at all, it's a few metal bars that get red hot. So yes, if you touch them you'll burn your fingers, but the bars aren't burning, they just convert the energy back from electricity into heat. Or an electric train: we've once before remarked on the extraordinary fact that all it takes to get an electric train moving is a power cable and a pick up. And what's so brilliant is that whether we want to light an electric fire in our living room or drive a train from Hamburg to Milan, the way in which we now 'package' or 'handle' the energy is the same: electricity can do either, and a million things on top.

That's why electricity is energy once or twice or more often removed from its source. The source can be anything. If it's coal, then you burn it to generate heat: the heat is energy once removed from coal. Turn that heat into steam and power a turbine, you get motion; the motion is energy once removed from fire and twice removed from coal. Neither the fire nor the motion though strike us as particularly abstract, and the reason they don't strike us as very abstract is because they're still visible, tangible, and there are only so many things you can do with either.

But the moment you remove the energy once more and turn it into electricity, you're really doing something new and quite different with it. Because now you turn it into something that can be anything. A fire can't be 'anything'. A fire can only be hot and moderately bright. On its own, the fire can't refrigerate your milk, and it can't store your photographs. It can't suck the dust out of your carpet, and it can't give you the football scores. It can't get you to watch the moon landing, and it can't calculate the interest on your loan. Without somebody standing by and making smoke signals or blocking it off and then revealing it again at specific intervals it can't even send you a short message. The most it can do, by burning, say, on a mountain top, is signal to somebody on another mountain top. That's about as far as it goes.

So it's true to say that we have in fact imbued energy with meaning for as long as we've been using it. But up until the moment when we start to use electricity, the potential uses that we could put energy to, and therefore also the potential *meanings* we could associate with it, were determined by its own 'type' or 'form'. So, much as there is indeed a range of uses we have for fire and a scope for forms that fire can take and the meanings it can have, they are all limited and very much defined by the nature of fire. Once we turn the fire into electricity though, the range of uses and the scope of meanings becomes *unlimited.* We no longer need to talk of potential uses and meanings, we can simply talk of *potentiality* itself, because many, you could say most, of the uses and meanings that we use electricity for could not even be imagined at the time electricity was first experimented with.

We can express this a little differently. We can say that up until about now, over the last fifty, sixty years or so, when we began to really understand and make use of digital encoding and computing, the application of energy was always derived from its own 'meaning'. 'Meaning' to us, of course, but nevertheless 'meaning' that was *inherent* in the energy itself. To stick with fire, because it is so elemental: the meaning that fire has to us is intrinsically linked to its own natural qualities and the way we experience them. We experience fire as 'hot', so it follows that fire to us means 'warmth, comfort'. We experience that fire scares wild animals, so to us it means 'safety'. We experience fire as lighting up the dark, so to us it means 'light'. From these meanings we can derive uses that again follow 'naturally'. Once you realise that cooked food tastes better and is more easily digestible than raw food, it makes sense to use fire to prepare food by cooking it. So close is the connection between the meaning and the use, that we consider them practically inseparable, and that too is no coincidence, because we most likely discovered the use by applying fire in its meaningful way, even if it was purely by accident.

With electricity, we are now able to put the energy to uses for which we may not even have a meaning to start with. The meaning may yet be invented or reveal itself. If you struggle with the thought of energy having any 'meaning' at all, you may find it easier to think in terms of 'form'. With electricity, it is no longer the case that the form that energy takes – such as fire – determines how we codify it. Instead, the way we codify energy – very specifically electricity – determines its application. You could say that the application is another, new 'form' of the energy. And so you have, for the first time, a *reverse* relationship between energy, its form and its use. No longer does the form of the energy determine its uses, instead the use of the energy determines its form.

And it is in the word 'codify' that you get a clue as to what we we're getting at here: looked at in this way (and we're aware this is one particular way of looking at it, this is not an absolute 'truth', nor is it an incontestable finding), energy could be said to be *symbolic.*

Again, it doesn't come so easily to us to think of energy as 'symbolic'. How can energy be 'symbolic', you may ask, and it's certainly true: the way we *experience* energy is not as 'symbolic'. But that's precisely the point. The way we experience energy is not 'symbolic', because to us energy is *'real'.* And that's because whenever we come in contact with it, we do so when it has taken on a form that we can relate to again: heat, motion, light. We *always* experience energy as 'real' and not as 'symbolic', because there is no other way for it to manifest itself to us. Before we turn it into electricity it is either heat or motion or light or sound, and when we become aware of it again, we do so because we've turned it back into heat or motion or light or sound. And these are not abstractions nor are they symbols of energy, they are, to us, *realities.* So what about electricity? Is that not 'real'? Of course it's real enough, otherwise none of this would be happening, we wouldn't be writing, you wouldn't be reading, there would be nothing on TV tonight, and nobody would be talking about an energy crisis. But how would you describe it, as what? Well, exactly.

X   ENERGY AND SYMBOLS

We started using symbols from the moment we took our first steps towards civilisation. In fact you could argue, and we do, that for anything resembling civilisation to exist, you need these two things: energy and symbolisation.

And why is that? Because without energy, nothing goes. The amount of effort and time we would have to spend on just surviving (for a very short lifespan, it has to be said, and none too pleasant at that) would be such that cultural conduct would be practically impossible. Everything we do that we associate with leading a civilised existence, from just boiling water for a safe drink to clothing ourselves, right through to mapping the human genome and finding treatments for cancer, rests on our ability to utilise energy. And without symbols nothing makes sense.

If you sit in a clearing of the forest and you hear the whisper in the trees, you can either just accept that that's what trees do and not ask any questions about it, or you can wonder 'what's that sound?', 'where does it come from?' and 'why is it there?'... The moment you ask these questions – and as conscious beings we are, as any parent of a four-year-old will confirm, apt to ask them – you need to find answers. And if the answers are not readily available, you have to devise possible answers; answers that are plausible enough, at least for the time being, to give you a sense of certainty, of stability. It doesn't, to this end, even matter so much what the answers are, or the concepts that you employ to come up with them. Before you have any concept, for example, of climate and differentials in air pressure that cause a breeze which in turn rustles the leaves of the trees, which causes sound waves to be carried through the air to your ear, you may have a very different set of concepts. You may think, for example, that the wind is a god.

Or you may decide that although the wind itself isn't a god, it is in fact a god's breath. You create a set of symbols. No longer is the rustle of the leaves something that just happens and that is meaningless and that you can ignore because it's always been there and will always be there and who are you to question it; now the rustle of the leaves stands for something. And something stands for the rustle of the leaves.

As soon as you have the ability to invest objects or phenomena with meaning, you can devise meaning with a purpose: you can begin to define symbols that mean something to a particular end, symbols whose meaning is recognised, adopted and developed by your community, and that thus stand the test of time. The symbols become useful, and they become a way of communicating meaning among you and your fellow human beings, not just today, but across the ages. This means you can plan: you can make arrangements for the present *and* for the future, you can make contracts that you can pass down the generations. Among the earliest civilisations, for example, families and tribes, when visiting each other, would take a piece of ceramic, wood, stone or brick and break it in two, each party keeping one half of the tablet as a symbol of their relationship, their loyalty, if you will, to each other.

The very word 'symbol' comes from the Greek 'symbolon', which literally means 'thrown together' or 'put together': this is thought to possibly refer to the two halves of such tablets being reunited to 'symbolise' (in that case 'put or bring [back] together') the meaning, here of the relationship or loyalty.

(Interestingly enough, the first recorded use of the word 'symbolon' appears in the Homeric Hymn to Hermes,[CXCIII] one of 33 hymns that were either written by, or in the style of, Homer, each celebrating an individual deity. In it, Hermes, the Greek messenger god, spots a tortoise and exclaims: "A symbol of great luck to me so soon! [...] With joy I meet you!" He then picks up the creature and promptly turns it into a lyre, thus going down in mythology as the inventor of said instrument, though not necessarily as the kind of god you can trust with your pet tortoise...)

Perhaps a little example: say you put on a remote island in the middle of the ocean two shipwrecked sailors, and they encounter there a fruit neither of them has ever seen or tasted before but that is delicious, nutritious and in plentiful supply. They will soon make up a word for it. The wongle fruit, as they might call it, because it might remind them of a wongle, which they'll have a good laugh about, because only they will know what a wongle is, will soon acquire its own connotations and associations, and it will become part of their culture. Wongle juice, sea bass grilled and served with wongle, dried wongle and wongle powder will not be far off. And even if the two stranded sailors do not speak the same language and do not come from the same background, they are still capable of naming their fruit wongle or anything else they like the sound of, and the word may still become part of their little island culture. Until they are rescued by a passing vessel, take some wongle fruit with them and start an immensely successful wongle business, soon exporting wongle to every corner of the world, patenting their wongle soft drink and turning Wongle into a trademark and global brand. Because of its great health benefits, their ethical farming practices and a clever, sustained marketing campaign, Wongle may soon stand for something that is universally good and beneficial, and the word may become part of people's vocabulary around the world. They may start using it to express something really wholly to be endorsed, and say things like 'he's a wongle of a man', or 'we've had a wongle holiday'... It may even turn into a verb: 'I don't know how she did it, but she's certainly wongled it!' The humble wongle has become well and truly symbolised.

Symbols have been created, collected and traded for millennia in all kinds of contexts, not just economical and legal, but very significantly also religious or mythical. Since time memorial, symbols have served us as an expression of inner bonds, as manifestations of relationships and significance: from our earliest ancestors with their friendship plates, to kids today swapping friendship bracelets in the school yard.

But what happens when energy – one pillar of our civilisation – is brought together with symbolisation – another? Before we get to that level, we'll have to take in two more steps in the progression of symbolisation.

(We'd have to go into quite a bit of detail if we wanted to do this justice fully, but we should perhaps bear in mind also that this is a book about energy. So we are sure you will forgive us if we give you the abridged version here: we hope that the dots are sufficient, and sufficiently close to each other, for you to be able to connect them into a coherent story that you can follow right through to the end where we'll be right back on the subject of energy.)

With our ability to use language it was perhaps only a matter of time before we would want to find a way of representing language in a way other than by speaking it. The fact that we are now finding this progression somewhat obvious does not take away from the fact that it was a huge step: the invention of symbols for words has propelled us forward in a way only few developments have, one of them you could say being the ability to handle fire, one settling, one inventing the wheel.

Once you can write something down, you can fix it in time: you can record it. Both for reference and for posterity. The actual meaning of your symbol no longer has to rely on tradition or hearsay, you can nail it. So the gesture of loyalty, for example, of our symbolon above, can now grow into a fully-fledged contract. With terms and conditions.

When we began to use literal symbols – signs and characters that were able to represent material things such as objects, buildings, animals or people – symbolisation reached a next level: we nudged one step closer towards abstraction, from situation or myth, you could say, to *concept*.

The word 'chair', for example, symbolises a thing that is made so we can sit on it. The 'bond of meaning', so to speak, that connects the word 'chair' and the thing that we can sit on passes from all similarly shaped or intended objects to all people who share the ability to interpret the word. And the ability to interpret it can, as we all know, be taught and learnt, both from scratch (from a position where you have no word for a thing that you can sit on) and from a comparable code (from a position where you have the word 'sedia' for an object you can sit on, for example, because your language happens to be Italian). Similarly, the character '椅' symbolises the word 'chair' that in turn symbolises the thing that is made so we can sit on it, as well as the thing that we can sit on itself.

Of course, over time our use of symbols and the way we think about them has changed dramatically. Early symbols tended to be imbued with a spiritual or mythical dimension. They obtained their validity from a shared *belief*. If everybody agrees that the sun is a god or that a person suffering from sporadic, intensive fits is possessed of an evil spirit, then that's perfectly valid in the context of that society. As long as everybody *believes* that slaughtering a lamb will appease the gods and bring good fortune to the people of the town, then that's what slaughtering a lamb does. The iteration, reiteration and active sharing of the belief reinforces and empowers the symbols that embody it.

Science (or, to be precise, what we today call 'science') changes our relationship with symbols. Now, symbols are relieved from their mythical or spiritual quality and they become representations and notations of what we consider pure, material 'fact'. From this secular position, the sun is no longer a god, it's a star. The person having a fit is no longer possessed, they are epileptic. Slaughtering lambs does nothing to appease anyone, it's just a waste of a perfectly good sheep. Where before there was a shared belief, there now is a shared *standard.* The reason a metre is exactly a metre is not because the spirits have whispered it or because god has ordained it thus, but because everybody agreed to call a tenth of a millionth of the distance from the equator to the North Pole a metre.[CXCIV]

The meaning is, as much as it can be, *objective,* and in this lies rooted our very strong sense that anything that is based in nature and that is scientifically valid – for example because it works – is by definition 'true'. And since we value 'truth' highly, it is also, by extension, deemed 'good'. This marks another step of abstraction, from *concept* to *operation.* Because now not only do we have a way of symbolising abstract notions (a metre is a purely abstract notion of distance), but we are putting it into a form that *works.* And it always works, because it's 'scientifically *proven',* which means nothing more and nothing less than that it is conceptually coherent and practically applicable, and that the form of abstraction has been formally generalised: a metre is *always* a metre, and exactly a metre, whether you're in Paris, France, Paris, Texas, or on a plane anywhere in between. And the reason you can have a plane is exactly that this is so. It may still amaze you, every time you sit in an aeroplane as it races along the runway, that eventually it takes off. But it doesn't *surprise* you, because the science is sound: planes that are built in the right way will fly, that's what they do.

From the advent of science up to this point we are in a world that we call *rational.* By this, we generally mean that things stand in a defined, specific relationship to each other, which only alters when the conditions or the things themselves alter. We tend to like rationality, because we experience it as reliable. If I build a bench with a plank of wood, then that wood will carry a certain amount of weight. And every time I sit down on it, I will be sitting comfortably, as long as none of the parts of the 'equation' change. If I leave the bench out in bad weather for a long time, the wood will eventually rot and be less strong, so when I sit down on it, it may break. Or I may put on weight and double in size, through eating bad food. Or I might sit down on the bench together with my uncle Tim, who could do with losing a stone or two. In these instances, I may be prepared for the eventuality of the bench breaking. But on its own and without any external factors being altered, the bench is not going to suddenly fall apart. If it did, I would consider that very strange indeed. And highly irrational.

But about halfway through the 19th century something happened which changed all that. All of a sudden, tangible, material science, the kind of science you can measure and touch, was brought into question. This shook things up so badly that there was talk of a fundamental crisis in mathematics, where it expressed itself in the guise of a 'new' type of numbers which had characteristics that seemed baffling. And that is why at this point in our proceedings we want to go one level deeper still and also talk a bit about numbers. It will take us right to the essence of what we are talking about, and help us understand why and how such big shifts as the one we say we are experiencing in the context of energy today are even possible. Because there is really nothing magical or mystical about it, it can all be traced right back to where we are at in our ability to think and our ability to put that thinking into practice and apply it in our everyday lives.

XI   NUMBERS AND ELECTRICITY

The first thing to note about numbers is that they are inventions. Our instinct is to reject this idea outright and say, of course they're *not* inventions, they are objective truths. But this is just because that's what we're used to: we've learnt them and developed them and that's why to us they now seem the only logical, possible way of expressing things. Yet, they are still, in essence, agreements. Like concepts, or words, they are man-made constructs. And as such they keep evolving. We, ourselves, keep evolving and developing our understanding of numbers, and we keep inventing new numbers, new *types* of numbers, to handle the evolved thinking we do. None of this means that they are arbitrary, though. This is important to emphasise: what we're making up is not that which we describe with numbers, what we're inventing is the code, the language, so to speak, that we use to describe it, it's the numbers themselves.

When we look at the different types of number we know, and how they relate to our understanding of the world, we immediately realise how obvious is this fact: that numbers are our own invention to express things that are not arbitrary.

NATURAL AND RATIONAL NUMBERS   The type of number we are most comfortable and most familiar with, and this hardly comes as a big surprise, is the *natural number.* Even the name we've given it, 'natural', suggests that it is a kind of number that just exists as it is, in nature, on its own. And in a way it does. Natural numbers are those we can count on our fingers, 1, 2, 3, 4, 5, into infinity. They are whole numbers and they are positive, which is exactly why they are so natural to us. When we go into the garden and pick up the apples that have fallen off the tree, and lay them out on a table, we can see and touch and sense very clearly that there are so many apples as there are, no fewer and no more.

For your most basic mathematical needs, natural numbers absolutely suffice. You can give away some of your apples, and thus have fewer than you started out with, or your ten-year-old can climb the tree and get some more apples directly off it, and you end up with more. No sooner do you start doing anything more interesting with these apples though, than your natural numbers are no longer enough.

Say your tree is a bit meagre and has only yielded two apples so far, but there are four of you? In practice, that's no problem, because all you do is take a knife and cut these two apples in two and everybody can have a half. In mathematical terms, though, you've already reached your first set of limits and you *have to* make up a new kind of number to deal with this situation, even though it is far from extraordinary.

The way we solved this was by thinking in terms of fractions with numerators and denominators: a number that tells us how many parts of a previous whole there are and also how big those parts are, in relation to the whole. So in this simplest of examples in which you, your ten-year-old Ben, your lovely wife Berta and your good old pal Paul each get half an apple, we most conveniently have 4/2 or four halves. This still strikes us as very normal and ordinary, but already we sense that we're not leaving nature untouched, that we are bringing in reason to work with nature, and handle it. So although the term stems not from 'reason' but from 'ratio' – the ratio between the two numbers that make up the fraction – it is apt that we call these numbers *rational.* We are now able to deal with objects and quantities on a much more elaborate level; we can calculate proportions and shares of entities, which means we can work out interest, for example, because we can relate an amount of money to a proportional length of time. We can also make calculations with geometrical forms and shapes, and so our architecture does not have to stay empirical: we don't have to just see how many bricks we can stack on top of each other before the wall collapses, we can project and plan and make accurate predictions as to how quantities, mass, and structural relationships will behave.

At the same time, we begin to understand and formulate, and work with, mathematical patterns. For example, we not only recognise the symmetry of an equation such as $3 \times 3 + 4 \times 4 = 5 \times 5$, but we find that our logic allows us to apply this regularity as a principle. So it makes sense to us to find that $33 \times 33 + 44 \times 44 = 55 \times 55$, and we can use this insight to make statements about what is *always* the case. We can be certain and provide logical proof that *any* similar pattern in which numbers consisting of the digit 3 are squared and then added to numbers consisting of the same magnitude of the digit 4 squared, yields a result that equals a number consisting of the same magnitude of digits 5 squared. And we can work out that this is also the case if we use multiples of the set of figures we've used. So $6 \times 6 + 8 \times 8 = 10 \times 10$. But the same is not true if we alter the pattern: $6 \times 6 + 9 \times 9 = 117$ and that does not fit anywhere because the result is not a natural number squared.

If you think that this is all sounding a bit obvious and banal, then you are instantly forgiven. Not because it is, but because to us it looks like it is, we're so very familiar with it all. But it's worth bearing in mind that up until this point, there are still two major components missing that actually make up what today we call rational numbers: negative numbers and zero.

Considering how commonplace they seem to us, it's astonishing to think that for hundreds of years, people who pioneered mathematics and geometry, and who were perfectly capable of working out highly sophisticated calculations, had neither negative numbers nor zero. For the Ancient Greeks, the idea that you could subtract 5 apples from 3 apples didn't make sense. You could add, and you could subtract, and you could divide and multiply, as long as you were left with a positive number, even a fraction. But it would not make sense to think of what you could do with minus two apples, since there was patently nothing you could do with apples you didn't have. And not only did you not have them, you had a supposed quantity of not having them, which was absurd.

The Greek mathematician Diophantus,[CXCV] who worked in Alexandria in Egypt during the 3rd century CE, and who is thought of by many as the 'father of algebra' considered an equation that produces a negative result 'false', yet his even more famous countryman Euclid,[CXCVI] who also worked in Alexandria, but six hundred years before him, was still able to, effectively, invent geometry.

The concept of negative numbers is first documented in China some 200 years BCE, though it's possible that they were understood considerably earlier. By the 7th century CE, we know for certain that Indian mathematicians used negative numbers in their calculations, and around the same time, something absolutely marvellous happened, without which our culture, our economy, our technology and our science would not be possible: the zero was invented.[CXCVII]

The zero is something of a conceptual wonder. So much so that books have been written about it and entire TV series dedicated to it. It is in itself a delectable paradox, because it *is* nothing, but nothing now goes without it. The way *we* understand our world today, we can't think away the zero. Yet we didn't have it, for something like nine tenths of our history to date.

With negative numbers and zero, the 'set' of rational numbers is now complete. It consists of all whole positive and all whole negative numbers and zero (together they are what later was given the name 'integers') and all positive and negative fractions that can be made by any combination of any of these, except where zero is a denominator, because there is no place in our logic (yet) where 3/0 or three zeroths makes sense. But there is no reason to be certain that this will by necessity forever be so, if what happens next is anything to by...

REAL AND IRRATIONAL NUMBERS   The zero is so groundbreaking because it describes something that we've always known to exist, even though it is nothing, and therefore doesn't exist. And not only does it describe it, it allows us to work with it, and with it establish an entirely new way of thinking, all on the basis of something that isn't even there. With zero, we are able to operate a decimal system and work with percentages. We can not only quantify the absence of a quantity, we can build a whole logical construct around it. And we can have a new class of numbers: real numbers. And that includes numbers that are not rational.

Irrational numbers are numbers that cannot be expressed as a ratio between two integers, such as 1/2 or 2/3 or -4/5, but that can be expressed by decimal numbers whose numerals after the decimal point never end and never repeat. This makes them particularly fascinating, because at first glance it would appear that there are very few instances when this would be the case, and that therefore there would be very little need for such numbers. But you can see where this is going: nothing could be further from the truth. Not only would nothing that we associate with contemporary living be possible without irrational numbers, but almost all numbers are in fact irrational.

If we draw a line and plot on it all the numbers we have encountered so far, that is all rational numbers, we end up with a densely populated string that extends from zero into infinity towards plus and from zero into infinity towards minus. There is no point that we can think of at which it will not be possible to add another number, and therefore the number of rational numbers is not just vast, it's infinite. But dense as they are, these dots would not constitute an uninterrupted line, since every number would only mark one particular point on it and there are in fact points on the line which cannot be expressed by integers or any fraction made up of integers. These points would seem to be infrequent and highly specific, and while in the latter lies their great importance, the former is deceptive. Like all other numbers, irrational numbers are a concept we invented to describe a reality. They are made possible and necessary, both at the same time, by our understanding *of* that reality. We have irrational numbers because we need them and we 'get' them when we are ready for them, you could say.

Take the circle and π. Pi is possibly the most 'famous' irrational number, and even those of us who never paid the slightest attention in maths will remember that it is intrinsically linked to the circle. Now clearly, circles have been around for as long as humanity, and much longer. And it did not elude early mathematicians, long before they had the zero or a name for rational or irrational numbers, that a circle has a circumference that is roughly three times as long as its diameter. But the *exact* length cannot be expressed as a fraction. It can be *measured,* and an exact reading can be obtained, but it is impossible, even today, to give the *relationship* between the diameter and the circumference of a circle its exact value.

It's vexing and exhilarating in equal measure to think that one of the most fundamental shapes in our universe has built into it a relationship that we can *never* accurately express. We can come quite close. As of 2015, π had been worked out to some 12.3 trillion digits after the decimal point.[CXCVIII] This gives us a very close approximation to the value of Pi indeed. But it is not the absolute value. And because the number is irrational, there can't be an absolute value. The same is true of any irrational number. And although most of us will only ever have heard of two or three irrational numbers, π being one of them, √2 being another, the golden ratio φ (phi) another, there are an infinite number of irrational numbers. (And because rational numbers can be counted, whereas irrational numbers can't be counted, it follows that almost all real numbers are irrational.) [CXCVIX]

We are able to use rational and irrational numbers in combination with each other, even though they appear to sit side-by-side conceptually, and even though they lead to two totally different types of infinity: rational numbers perch on the unending line of numbers in the plus/minus direction where you can always add one more, whereas irrational numbers are like anchor points on that line, off which they branch into another direction (or even dimension), by leading into the unending line of never repeating numerals after the decimal point, which we can't simply add to but which we have to calculate, and which is therefore specific but unpredictable, but predictably unending.

That in itself, the idea that there is not just one infinity, but several, is one we didn't always understand, and we still today don't see it as obvious. But it is nevertheless borne out by mathematics, and therefore as real as an apple from your garden on your breakfast table. The point though is this: we open up new types of possibilities, potentials, by getting to grips with concepts that previously were simply out of reach. Mechanical machines and apparatus, steam engines and industrial processes, but also things like timetables and an ability to synchronise time over continents, seafaring, aviation, exploration, these are all impossible and literally *unthinkable* without the science behind them, without a mathematics that includes irrational numbers, zero and negative values, fractions and integers.

Each time we invent a new class of numbers we open up a new level of what is possible. And this same thing happens again, with yet another class of numbers, that are even more abstract and even less tangible, and just as much part of our universe...

COMPLEX NUMBERS   Complex numbers combine real numbers with *imaginary numbers.* In mathematical notation, an imaginary number is written as *bi,* where *b* stands for any real number that is not zero and i stands for the 'imaginary unit'. The imaginary unit is the square root of -1, which we would normally write as √-1. But we don't, because it is 'imaginary': as your calculator will tell you when you enter -1 or any other negative number and then hit the square root √ button, the result is *Error!* Square roots of negative values are not meant to 'exist'. Much as our friend Diophantus simply declared that any equation resulting in a negative value was 'false', so our understanding of maths leads us to declare that the square root of a negative value simply doesn't exist and is therefore wrong. And much as we've learnt to work with negative numbers, whether they 'exist' or not, we can also work with the square root of negative numbers, such as the imaginary unit √-1 that we call *i,* and any imaginary number that is made up of *i* and any real number *b.* It only took us a while to realise this.

If you combine an imaginary number *bi* with a new real number then you have a complex number *a + bi.* Calculations involving complex numbers are consequently called 'complex calculations'. What's peculiar, special and in no small measure puzzling about imaginary numbers, complex numbers and complex calculations is that they cannot be represented by any known method, because, as we've just seen, they do not, in that sense 'exist'.

And *that* is what caused the crisis in mathematics. Up until this point, we were able to express our mathematics in geometry. Being able to express mathematics in geometry meant being able to visualise it in a way we can readily grasp. Suddenly, though, it is now possible to work with units and in dimensions which previously could not be expressed. And although it is impossible to represent complex numbers in a way that we can see and easily comprehend, the calculations they enable, and by extension the effects they have, are absolutely tangible and real. And it is perhaps no coincidence, then, that electricity, which as we have seen has marked one of our greatest and most breathtaking advances in science, is impossible to work with and calculate without complex numbers.

It is complex mathematics that has opened up electricity, and therefore information technology, to us. And there's a fascinating parallel in that a type of mathematics based on a number which fazed mathematicians to the point where they gave it the name 'imaginary' is in turn the basis for a technology to which we have given the name 'virtual': it's there, we can work with it, we can make it do things and shape our world with it, but we can't make it tangible, we can't actually see or let alone touch it.

XII   POWER THROUGH THOUGHT

Our short expedition into the extraordinary world of numbers serves to illustrate that we make quantum leaps because we can. And we can, because our understanding, our intellectual capacity, matures over time.

Not by itself, but because we investigate, we labour for it, we think and challenge our thinking, we try and we fail and try again and prevail. It's a very human thing, abstraction.

With natural and rational numbers we are able to think through and work with Euclidian geometry. We can work out proportions and relationships and discern patterns and regularities. We know about π, but we can't as yet quantify it, we can only approximate it, by trying to square, very literally, the circle. Once we have real numbers, we can work not only with fractions, but also with percentages, we can employ irrational numbers and give them ever more precise values. We can begin to conceive of different types of infinity and we can use infinitesimals in our calculations, even though by their very nature they are not quantifiable.

With complex numbers we can make calculations that are so abstract that we have not, as yet, developed a way of visualising them: we can't really imagine the entities we're dealing with, but we can deal with them to the most spectacular effect, as for example in manipulating and using electricity and information technology.

Not just in science, but in the arts, in our social organisation, our take on religion and our understanding of ourselves in this world, in fact in every aspect of our culture does the way we think manifest itself. We have said earlier that two of the principal pillars (and we are not suggesting they are the only two) of what we understand as civilisation are energy and our capacity to symbolise. It is no accident then that a development in the level of symbolisation that massively impacts on energy should be of great significance to us. And what are numbers other than symbols. It is our ability to symbolise electricity and electronics, through the use of complex numbers, that allows us to build modern cities, to have the internet, to talk and text on mobile phones and watch television. Nobody *quite* knows what electricity is, but we're all happy to use it. Similarly, nobody is *quite* able to imagine the imaginary unit, but it works. We can't grab hold of electricity, but we can shift tons of steel with it. We can't draw or build, in any dimension we as yet understand, a visual or physical representation of a complex number, but we can make it illustrate anything we've ever seen or been able to imagine, on screen.

So if we keep saying that we are now in the era of potentiality, that the limitations of the material, resources-based world have become immaterial in terms of energy, then this is not based on a whim. It's based on a mathematical principle, and it is replicated in pure theory. So *of course* it's no coincidence that a mathematics that is completely removed from substance and geometry gives us access to a type of energy that is completely removed from substance and geometry. The two go hand in hand. And it is entirely possible, you may say likely, that at some point in the future we will discover yet another type of number, opening up yet another kind of mathematics with yet another infinity that gives us access to yet another kind of energy. But we can't conceive of it. Yet. Because we're all still learning.

You may remember that at some point, quite early in the book, we ventured that we were lagging about 135 years behind our own thinking when it comes to energy. For about this long we've known how to handle electricity and we've become really quite good at it. But it's only recently that we've started to recognise that with the mathematics and therefore the theory of electricity being so different to the mathematics and therefore the theory of any other form of energy, we get a completely new kind of potentiality: the virtual.

Up until just a few years ago, the majority, if perhaps not all, of the roughly five hundred billion electrical devices in use, either have been, or can be thought of has having been, conceived as instruments, meaning they each had a specific purpose as a tool, and with this purpose went a fairly robust but also inflexible method by which they function properly. A power drill, for example, can only be handled safely in so many ways. So from the outset it looks as if that's all these machines really are: mechanical instruments or tools. A washing machine would appear to do just exactly what washerwomen used to do, only it never tires and doesn't get rough skin.

Yet the technological principle behind these devices is not actually physical, it's *informational.* What makes a washing machine or a power drill work is electricity. And electricity is capable of far more than simply lending force to a mechanical process of, say, turning a drill bit or spinning a laundry drum. If we think of a power drill or a washing machine as just a continuation of mechanical technology, what we are doing is completely short-changing the devices, the technology at their core and ourselves. Looking at a power drill as just a drill with more power than, say, a mechanical drill in the hands of our mate Ed, is really inadequate. Of course, a drill is a drill is, after all, still a drill. But as a *concept* a power drill is not just tied into the world of physical forces and mechanics that we have learnt to understand so well over the last couple of thousand years, it is in actual fact tied into the world of electric currents, which means it is tied into an *abstraction* of physical forces and mechanics.

As an abstraction of physical and natural forces, not only is electricity capable of delivering power to the machine (a set of wheels, gears and transmissions could do the same thing), but it is also capable of *controlling* it and in so doing of defining the physical forces it supplies power to: it is only a matter of design whether our power drill actually turns a drill bit, or whether it turns the drill bit and shines a light on the hole in the wall to be drilled, or does both and play a tune by Muse at the same time, or whether it does none of the above but works out the monthly repayments on your mortgage instead.

No amount of mechanical engineering can make it do that. But about sixty cents worth of hardware and a bit of programming can make it do that very easily. You may query the 'use' of a power drill that does precise calculations but can't make a hole in the wall, but that's just a question of how you define 'use', 'drill' and 'hole in the wall'. In the world of mechanics, apparatus and physical instruments it may make sense to take a narrow view on such things. But the moment you build an instrument (like a power drill) around the abstraction of energy that is electricity, you turn the instrument into a medium that can run an application, which means it can be controlled any which way you like, it just depends entirely on what kind of a system you want the device to belong to, how you want it to relate to the system and all the other devices in it, what, therefore, in short, you want your device – whatever you then call it and whatever type of functionality or functionalities you associate with it – want to *be*.

Anything that 'is' has to be so in a way that is compatible with everything else that forms part of the same system. So if the system the power drill in our example belongs to is a normal household that has about two hundred other electrical devices, that has a mains power inlet and possibly even a way of feeding power back into the grid, then anything the drill does has to not limit or adversely impact on the behaviour of any of the other devices. A power drill that, every time you turn it on, blows a fuse, for example, would be annoying to the point of being useless. But an 'intelligent' device can go far beyond not blowing a fuse. As we have seen, an intelligent device can act like an application: it can do certain things in a certain way that is relevant to something else that is happening around it. There's no reason why you shouldn't be able to have a drill that checks for wires in the wall and tells the tumble dryer not to kick in for a few minutes; or windows that close and lock when you leave by the front door while there's nobody else at home. Or a TV that pauses when the phone rings. We don't really know what uses there are, but if you can think of one, then with this technology, you can probably do it.

The implications this has are immense. We have developed all manner of these 'abstract machines' that are the great-grandchildren of Turing's 'universal machines' and that are capable of running all manner of applications that we use to record, store, convert and distribute symbolic processes, which, in their entirety, we call 'media'. And they play a critically important role. As media proliferate and penetrate ever more deeply ever more layers of our daily lives, more and more – ultimately all – of the actions, processes, traditions, institutions, rituals and codes of conduct that we use to manage, structure and organise our reality become open to individual user input and to the level of mutual exchange that this entails. We can see this played out every day on the internet where blogs, tweets, status updates, and video, picture and text content that is generated, commented on, rated and shared by users does not play a side show to the traditional, and traditionally controlled, media any more, but takes centre stage in shaping behaviours and directly influences major outcomes, such as that of general elections, the actions and decisions of global corporations and the direction our culture is heading in. And as we have seen, this same principle now also applies to energy.

Once you treat energy not as a resource or as something that you have to laboriously generate and then, carefully measured and costed, distribute or disperse, but instead consider it much in the same way you would treat an email that you send from one computer to another, casually channelling it through the system to precisely where it is needed, in the exact quantity desired, you realise that energy has to be no different, really, to data: 'I have some here, I can send it to you right away.' We make nothing of it, when we send a bit of information by email over the internet or by text via the mobile network. So why should we consider it strange if we can also send energy over the power network?

XIV   POPULATIONS AND MODELS

We don't mean to overburden things, but neither do we want to take any unnecessary shortcuts. And so there is one more aspect we'd like to consider for a moment.

One of the things that excites us about the energy future we envisage is, as you are aware by now, its potentiality. And much of this potentiality lies in the fact that when you are dealing with something – anything really – at an abstract, symbolic, level, you can not only imagine it in ways that you are able to imagine now, you are able to imagine it in ways you can't even imagine yet. And though that sounds paradoxical, it makes sense, because you have no idea what is going to be possible, so how would you imagine it, but you know that it is going to be possible. You have to allow for the fact that your potential far exceeds your current capacity to make use of it. We argue that the kind of symbolisation made possible by digital code opens up a new game on a new plateau.

Now, science is our means of connecting the dots between phenomena that we can observe. If I'm lying in the grass looking at the sky and suddenly there's an apple falling from the tree, I witness a phenomenon in action: gravity. I may not at first have a name for it, and I may not immediately know how it works, but if it prods my curiosity and I want to understand it, I can set about making other objects fall to the ground from different heights in different settings, and I can measure, note down and systematically observe the behaviour of these objects as they fall to the ground. Before long I will have a set of observations which show up certain regularities which allow me to make predictions about how things fall to the ground. I can now make calculations and postulate rules and formulas about objects falling to the ground generally, and I can make reasonable assumptions about a force they are subject to, which I can also give a name to, such as 'gravity'. I can test my assumptions about it by carrying out further observations and experiments and if I'm 'right', then the assumptions I've made will stand this test and will therefore be applicable and 'true'.

What I'm actually doing is creating a geometrical representation (a formula or calculation in space and time) of the phenomenon I've observed (the instance of an object falling to the ground), and then transform this geometrical representation into analytical functions (establish a rule that applies to *all* instances of objects falling to the ground), as well as set up chains of cause and effect (work out what happens when instead of an apple falling to the ground I have a constant flow of water hitting a turbine from a certain height, giving it a certain speed, and using the motion to power a millstone, for example; in other words define the relationship between individual instances of different phenomena that I can observe in the world).

It is this way of understanding the natural world, which we are so very used to and which we may call a rational, geometrical science that is in essence empirical, based on things we can observe and measure, that is being augmented by something new. And what it's being augmented by – you could say what it's dissolving, or, if you like the term, evolving into – is a science that's based on *population dynamics* and *probabilistic modelling.*

POPULATION DYNAMICS   Now, population dynamics are pretty much what it says on the tin: they describe how 'populations' behave and how their behaviour impacts on individual members of the population as well as on other populations. A 'population' here can be anything that consists of one or more individual members that are capable of action and/or reaction. Normally a population can grow by members joining it, and decrease by members leaving it. Populations may be made up of people, but they can just as easily be made up of animals, bacteria, or cells, or virtual entities, such as computers on a network, or individual units of data on a computer, or clusters of electrons.

What makes populations interesting is that they are never static and that their individual members *behave:* they aren't just there, they do things and what they do may be unique and specific to them or it may be in tune with other members of the population, and in either case it may or may not affect other members of the population, severally or jointly, as indeed it will, intentionally or not, affect the overall condition and behaviour of the population as a whole. Populations are, therefore, *dynamic.*

Say you have a village with a hundred people living in it. Assuming you start off with a standard mix of ages and genders, within a year's time you'll have lost a few of the old ones and about twenty babies will have been born. Some of the more adventurous villagers may have gone for a wander and discovered there's another village with two hundred more people, just a few hours' walk away. Depending on how friendly or hostile the other villagers are, they may kill the visitors, or invite them in for a brew. Friendships may form, even alliances. Cross-village marriages. A regular exchange may start up between the villages, and a road be built. Soon you may have people from village A settling in village B and vice versa. The two villages may grow into towns, and, as they expand ever further in each other's direction, meld into one big city...

So of course, population dynamics are nothing new in themselves. The reason they are becoming so important is because thanks to electricity, information technology and communication networks, more and more of our science can – in fact has to – be understood in terms of how populations behave, because we, and the things we create and observe, interact so fast and so intensively with one another. And things interact with each other that used to be either static or directional. Take road traffic, for example. It used to be a case of motorists sitting in their cars, getting stuck in tailbacks, and the local radio station, receiving reports from the police or from their own chap in a chopper about the tailbacks, putting this information on air and then watching as nothing much happened until the jam unclogged itself anyway. Now, mobile phone and GPS networks are working in partnerships to obtain realtime positions and flow patterns from cars on the road, feeding that into their system and the system presenting densities and forecasts about delays even before they occur. The driver listens to or sees on their display information that has been put together with information their own vehicle sent to the pool.

It's as if, in a way, we are no longer lying in the grass watching the apple fall from the tree. We're there to catch the apple and throw it right back in the air. And because everybody else is doing the same thing, there are suddenly a whole lot of apples flying around left, right and centre. We can't contend ourselves, any more, with measuring and understanding how apples fall to the ground, we have to learn to understand how apples fly across into the neighbour's garden, who they are caught by, whom they are thrown to, who they belong to before they're thrown, what they are doing while they're in the air, what can be done with them once they are caught, what happens to the apples that end up on the ground anyway, and what the point is of apples that keep being chucked around instead of being eaten or made into cider.

And so the principles of universally applicable laws and formulas, as we know them from traditional science and mathematics, no longer dominate our understanding of the world. Instead of being able to say 'an apple always falls like this', the best we can do is say: 'as far as we can tell, most apples seem to behave like this, but they can also behave very differently, it just depends...' We can't even say 'an apple is useful, because I can eat it, make apple pie with it or turn it into cider' any more. An apple may or may not be useful, it doesn't matter. 'Usefulness' in itself becomes meaningless, because there is nobody to say that an apple that just *is,* for no particular purpose, can't be a wonderful thing, or that an apple which serves as a paperweight is any less useful at being an apple than a paperweight is at being a paperweight (or indeed at being an apple). And none of this is restricted to human level interactions. We are able to observe the whole universe in this way, and what we see is boundaries getting diffuse and delineations disappearing, be it in microbiology or quantum mechanics, be it in physics, chemistry or even astronomy.

This being the case, we have, it would seem, lost a great deal of certainty about our world. You sometimes hear people say that things used to be a lot simpler in the olden days, and while part of that may be pure nostalgia, there is some truth in the matter: our world – a world that is described by complex mathematics, remember – is an inherently complex one. But the world itself hasn't changed. *We* have changed, and with us our understanding of the world; our language and our science for it, and very importantly our ability to look for, interpret and understand the processes at work in it have changed. But the way we used to describe, understand and look at the world in the past was no more 'real' than what we do now, nor was it any less so. It's just that the grid, the rule, the hard and fast certainties of only sixty, seventy years ago, are not enough to describe the world as we see and experience it any longer. All these things going on everywhere at the same time are eminently unpredictable. What we have today is no longer just a tree and an apple and gravity and perhaps a bit of an easterly wind. We have all these additional individual forces and directions and trajectories and intentions working with, against, and on each other. So we can no longer say: 'If the branch is X yards from the ground and the apple weighs Y ounces and there is a light, force-2 breeze of between 5 to 7 miles an hour, then the apple will land with Z impact on spot A'. In this new situation, who knows where the apple will land? If Mr Thorncombe, who's a dreadful catch and even worse a shot, gets to it, it may never make it across the fence. If on the other hand his son Johnny gets a hold of it, it may land over at Number 22. Unless Johnny's been out drinking with his mates all night, and is now lying crashed out on his bed, in which case the apple will stay right where it is...

PROBABILISTIC MODELLING   This is where probabilistic modelling comes into it. Probabilistic modelling tries to make predictions not on the basis of geometrical calculations which stem from the observation of natural phenomena (the formula, in this case, that would describe the fall of the apple), but on behaviours of the same or comparable populations.

If you've been watching a neighbourhood where the residents keep throwing apples to each other for a while, you can start making some predictions about how the apples are *likely* to fly in the next period that is similar to the one you've been watching. There is no real *certainty* about it, because you have no control over the individual factors that affect the apples' behaviour (you can't know whether it's Mr Thorncombe who catches the apple next or his son Johnny, or whether Johnny's hung over or not), but you can, over time, discern patterns. You can establish, perhaps, that Johnny has a tendency to be on top form during the week but rather the worse for wear on a Sunday morning. But on Sunday afternoon, every second week, his cousin Gemma comes over for tea, and she's got terrific aim. So if you watch your population closely enough and long enough, you may realise that in actual fact your predictions overall are pretty reliable. You may notice alterations and developments in the patterns you observe and you may identify the probability of certain occurrences to be so high as to be almost certain or so low as to effectively render them random. You may work out reliably that between about midnight and six o'clock in the morning hardly anybody ever throws any apples at all. Or that of all the apples thrown, at least ten percent end up rotting in the grass. Furthermore, you may know some of the parameters in which apples are likely to be thrown. You may have some degree of certainty, for example, that no Olympic hammer throwing champion ever gets his hands on an apple here. And that any apple ever thrown will not be heavier than, say, half a pound. And so even though you seem to be dealing with chaos, you can in fact begin to draw fairly safe conclusions that are perfectly sufficient to build on: you can *model* behaviours based on probability.

These are still models, and as such they are not one hundred percent accurate, they are, after all, dealing with factors that are inherently unpredictable, but they are incredibly useful, because they allow us to play around with reality. 'Play around' may sound like a thing of little consequence, but in fact it's the opposite. In the context of, say, weather forecasts, early warning systems for tsunamis, earthquake prediction, town planning, air traffic control, transport, immigration or education policy, to name but a few examples, models become a matter of life and death, of positive outcomes versus very dire ones.

If we can model a possible reality, we can imagine it *with consequence:* we can look at not only how something works, but what the effect will be of it working in a particular way. And that, again, opens up vast potential. It means we can try things out that we would not otherwise even be able to think of. Because not only can we look at things we can observe (for example a population of apple-throwing neighbours) and create a model of probability as to what their apple-throwing behaviour will be like over the next few hours, days, weeks, months or years, we can also look at the various factors and imagine all sorts of conditions, constellations and behaviours which we've never observed, but which may one day come or be brought about. (For example giant apples that are twelve inches in diameter but actually only weigh a few grammes and so float on for miles without ever touching the ground...)

Obviously then, the kinds of models we use, either to analyse and understand things we can observe, or to imagine things that we can't, may vary a great deal. It is really a case of choosing the appropriate model for any given task. But this applies to all science. If we want to plan something on a two-dimensional plane, then standard or Euclidian geometry will do just fine. In Euclidian geometry, if you draw two straight lines next to each other so that they are both at a right angle to a third line, and extend them into infinity, they will keep the same distance from each other for as long as they continue, which is into infinity, and they are therefore parallel. No sooner do you draw these same two lines on a sphere though (such as on planet earth), than you realise that Euclidian geometry no longer applies. The two lines, which a moment ago were parallel, now begin to curve towards each other until they intersect. If you think that's unlikely, take a globe which has longitude lines drawn on it running vertically across the surface and you see exactly that this is the case: the lines meet at the North and South Poles. So, clearly, if you are going to model something large enough on a sphere small enough, you are ill advised to use Euclidian geometry, as it will be wrong: you need to use spherical geometry.

When two thousand years ago we were very much tied to the world as we could see and touch it, two hundred years ago we were well into calculating things that existed nowhere other than 'on paper'. And in the last twenty years or so we have edged firmly into that which exists only on our computer screens.

The key to this is symbolisation. Once you can symbolise the world as you know it, make it purely abstract and remove it completely from the absolute reference points, let alone the tangible objects that you're traditionally used to, you can invent it any which way you like. And you can do so scientifically, with applicable, valid outcomes. The world really becomes your oyster. That is the immense power of symbolisation. And as we have seen, this is nothing new, it has always been thus. What is new is just our arrival, relatively recently, at a new level of symbolisation.

XV   LEARNING TO LOVE UNCERTAINTY

We are becoming more familiar with our 'new reality' – really our reality newly understood – day by day. We are no longer bemused by the fact that there are metaphorical apples flying around left right and centre. We are, without necessarily noticing it, getting in tune with a way of being that only a short while ago would have alienated us beyond endurance. It is absolutely no coincidence that the word 'random' entered the urban vocabulary as an adjective of mild approval around about the middle of the first decade of the 21st century. 'That's so *random*' suddenly started to mean 'that's kind of cool'. Nobody born after about 1985 would seriously expect there to be a structured, linear order to the way they consume their news, for instance. When our generation, who were born in the 1960s and 70s, grew up with radio, TV and daily newspapers that delivered selective, edited and timed content in specific ways, today we are far more likely to catch a story break on Twitter or Facebook, and then check with a news website, and maybe then turn on the rolling news channel. It doesn't mean we don't watch *Newsnight*. In fact the opposite, we may still go to our traditional purveyors for in-depth coverage and analysis, but we're now perfectly used to everything happening all at once.

We are living in a very much softer, more fluid world than only a quarter of a century ago. And that means we ourselves can afford to become softer, more flexible, more vulnerable, perhaps. And there is good evidence that we are. If you live in London you may have noticed how, over the last 30 years, the look, tone and feel of young men has changed. Back in the mid 1980s, your average gathering of males in their early to mid-20s was likely to be characterised by a palpable degree of aggression and posturing. Obviously, you'd find exceptions, and it's impossible to generalise, but the way young men talked to each other, behaved with each other, their physicality, was rougher, edgier, and more wary of each other. You did not see blokes in a West End bar or club hug or kiss each other, unless you went to a gay bar or club. Nor did you really see men cry. A generation later, and you can still find your edgy, rough and physically awkward youth. But your average young man has morphed into a mild-mannered, physically unselfconscious urbanite with few issues and little to prove, and much more in touch with his emotions, which he's far less afraid to show.

Tribal allegiances, our sense of belonging, our sense of ease at being who and what we are have all shifted. Again, it is hardly a coincidence that traditional, monolithic institutions like political parties, big corporations or the church struggle to convey any appeal to us. We're getting used to picking and choosing, to floating, to putting together our own cocktail of life. To have a bit of this and a bit of that. It goes right and deep into the essence of our being, our actions and lifestyles. We create our own space and time and see nothing strange in doing so, even when we are with other people at the same time. Our social codes of conduct, our understanding of society, our perception of ourselves in the world, they are all in flux. And we also witness the reaction to this: the resurgence, suddenly, of authoritarian ideologies and reactionary values in some layers (more perhaps than circles) of our societies. This great uncertainty and blurring of boundaries and identities is not universally welcome: some people are deeply uncomfortable with it and yearn for the old, rigid structures that once upon a time made them feel much more sure of who and what they were. And that, too, is to some degree perfectly understandable.

What it all comes down to, though, and what we've spent so much time trying to explain and illustrate, is that by embracing the potential that our current, digital technology offers us, by lifting our thinking from the earth-bound, practical level to the virtual, imaginary level, by taking full advantage of the power of symbolisation and utilising mediality, we can conceive of and devise realities now, which have hitherto existed at best in fairy tales. And we can make use of them, without ending up in fairy land. We are at the point, today, where we can do something no generation has been able to do before us: we can *surrender control* and gain power from doing so, we can immerse ourselves in our own narratives, as characters and authors simultaneously. We can be living instances of serious storytelling.

Our interest, as you'll have noticed, is not confined to energy. And it isn't confined to technology either. In fact, we think that what we are on the brink of is an age in which we can move *beyond* technology. But is it possible to go beyond technology? And what would that mean? How would we do it?

The wondrous thing about virtuality, and this is neither an exaggeration nor poetic use of language, is not so much that it allows us to imagine things that otherwise we couldn't, but that it allows us to play with them, make them 'real' in their own way, even if maybe at first 'only', or only ever, in the virtual world. Because this in turn makes it possible for us to think beyond technological 'solutions' and instead think in terms of a technological backdrop or landscape.

The term 'technological solution' itself suggests one particular approach, which is an object, or rather objective-oriented one that goes more or less in a logical line, the way we are used to: it's cold in my house, I need to make it warm, because if I don't make it warm, my children will get ill and I can't think straight and get on with my work. Problem. Solution: I turn on the heating. The heating technology can be very simple: a stove with a log fire in it, or rather sophisticated, a fully integrated air conditioning system that regulates room temperature throughout the year. Still, the purpose of the technology is pretty much one of 'solving a problem'. Technological thinking, most if not all of it, goes in that direction. Even when the problem isn't one that we would *necessarily* consider to be one, the workings of the approach are still pretty much the same. At the time when Leonardo da Vinci [CC] began to design flying machines, you could not really have said, in all seriousness, that not being able to fly was a major 'problem' for human beings. The fact that today we fly routinely has turned any *in*ability to fly caused by external reasons, such as the famous dust cloud over large parts of Europe in 2010, into a problem, no doubt. [CCI] But before the 20th Century, flying was not in that sense high on the agenda. Nevertheless, the approach that da Vinci took, along with everybody since who developed flying into an integral part of contemporary living, was a solution orientated one which treated not being able to fly as a flaw.

Today we are able to go beyond that. In a world where there is abundant energy, we do not have to think in terms of problems and solutions. We can really remove ourselves from being thus tied into what's necessary, and look instead at what is thinkable, imaginable, dreamuppable…

Thanks to virtuality and the pooling of our imaginations that networks permit, we can dream up things that are not even in that sense technological, but that are really just the application of that which is possible *or even that which isn't yet but one day may be.* Not because we *have to,* but because we *can.*

This in itself is a defining aspect of abundance. Once you're able to do things for their own sake, just because it would be interesting, or entertaining, or enjoyable to do so, you have not enough, you have *plenty*. And once you have that, technology as such moves, paradoxically, into the background. You can start taking it for granted, as something that's just there. You can invent *irrespective* of how applicable or not things are. The technology is no longer an instrument to get a result, it's simply an environment. Or, perhaps more accurately, it's a part of the environment. The environment as a whole becomes medial.

Think of Hannibal,[CCII] if you like, and his elephants, trekking across Europe and over the alps.[CCIII] Imagine going up to him and saying: 'Wouldn't it be wonderful if you could hold in your hand a map with a pointer that tells you always exactly where you are and what the landscape is around you, and what the weather is doing. You would then be able to get to Italy much quicker and more safely.' He wouldn't have taken you very seriously. 'Like what? Like a *magical* map?' 'Yes, but you can flick between aerial view, the way a bird would see it, and street view, the way you see it, and just the normal map view, that tells you where things are...' He would not have understood. And you'd have had an exceptionally hard time trying to explain how and why this could be possible. Imagine on the other hand going up to him and saying: 'Here use this, it'll make sure you won't get lost', and handing him your mobile. He wouldn't necessarily grasp the concept, but he'd know how to handle it within minutes. Would he understand the *technology?* Of course not. Would that matter? Not one jot.

And this is what's different. If instead you'd gone up to Hannibal and told him: 'How about a machine that takes you across the alps in the air, wouldn't that be convenient?' He may have thought you a dreamer, but the concept would not have been in the least bit alien to him, or difficult to explain. 'You mean like a ship, but one that flies in the air, like a bird or a dragon?' Exactly.

We are, in some areas of our lives, already going 'beyond technology'. The map that to Hannibal would have been magical is just one of several hundred thousand possible uses we put our mobile devices to. We don't give it a moment's thought, we just interact. But most importantly: much of what today we do as a matter of course would not only not have been possible a hundred years ago, it would not have been *imaginable.* This is the key change that really matters. Right up to the moment when we start using electricity, practically every technological advance is a step on the same plateau as we'd been on since the very early days. Yes we got stronger, yes we got faster, these machines made us travel at speeds no horse could outrun, and they thrashed the wheat and spun the yarn like never before, but they were nothing, in character, but amplifications of what we were already capable of. That changes with electricity. *Nobody* had watched their own foetus in their own womb before, alive and kicking. *Nobody* had transported an image, in real time, across many miles and made it appear on a little screen, together with sound. And while these are in themselves magnificent achievements, what *really* unlocked the potential of electricity is the digital revolution, which is barely a generation old.

Because of our naturally short-term view on things (it's a big effort for us to think beyond our own lifetime experience) we divide the advent of electricity and the beginning of the digital age into two separate events. But in the much broader scheme of things, when looked at from a perspective that takes in 10,000 years of human history, the period we're living through right now is simply the second phase of this genuinely amazing transition we've been going through since the tail end of the 1800s. The electric 'age' and the digital 'age' are not separate, they are part of the same shift. No wonder the 20th century was one of upheaval.

Technology, the instruments and mechanisms, the systems and processes we use to make our lives easier, better, more interesting, more meaningful and very often simply more entertaining, is purely an expression of our desire and indeed our ability to do just that. It's part of what makes us human. Sometimes we get frightened and start to feel that technology is alien to us, that it threatens to enslave us. But technology is not an organism. Nor is it owned or controlled by anyone. Technology is just what we use to enhance our experience of living. And as we and our science and intellect and our understanding of the world develop, so of course does the technology that we invent and develop. We are not about to make any predictions as to what will happen in the next 50, 500 or let alone 5,000 years. Some people would question whether we even have another 5,000 on the planet. We don't know the answer to this. But what we do have great confidence in is our ability to continue to create our own story, in this next chapter, and those that follow.

We have, with electricity, ubiquitous energy. We have, with the internet, ubiquitous information. We are entering – this much we think we can say – the age of ubiquitous technology. And once technology is everywhere, it becomes unnoticeable, a fact of life. We just use it, live with and in it. We, in a sense, become part of it and it of us. Because it's there. And yes, we would be lost without it. But with abundant energy, why would we be without it?

THE OUTLOOK: WHAT NEXT?

X

In May 2011, the United Nations published a report on its up to then most expansive study into 'renewable' energies. We don't, as you will know by now, particularly like the term 'renewable', since we think of our principal source of energy not as 'renewable', but as infinite. (This, we realise, is in itself not absolutely accurate, since the sun will not last into infinity. In about 4 billion years' time or so, it will burn out. We will not be here, as a species, on this planet, to witness that event though, so to all intents and purposes on a human scale on earth, it is, more or less, infinite...)

The report, produced by the Intergovernmental Panel on Climate Change (IPCC) was quoted by Reuters as saying: "*The cost of most renewable energy technologies has declined, and significant additional technical advancements are expected... Further cost reductions are expected, resulting in greater potential for climate change mitigation and reducing the need for policy measures to ensure rapid deployment.*" [CCIV]

This is precisely what we think too. And since that report came out, its predictions have proved entirely correct, and they will continue to be so for some time. What lies ahead of us is change driven by economics. And that is just as well. We don't want to get into a debate about politics any more than we want to be sucked into discussions conducted on terms of ideology. There may be any number of causes for changing any number of things in the world, but they are not the concern of this particular book. Which is why we are satisfied to accept that in the world as it is structured and organised today, the *only* significant factor that can determine, in the long term, whether or not a set of technologies is going to be successful is economics. This doesn't have to be *simple* economics. Of course, we know that if something is cheap enough, or better still, *cheaper* to use than other technologies, then it will push through. But we have also seen that there are *qualities* that matter. We've looked, in some detail, at how the value we attach to something is not determined purely by what it is, but also by how we perceive it. We know that it is not a rampant 'survival of the fittest' type market that gives society best value. And we know that 'value' itself is a relative term that is defined by culture, by tradition, by history, by personal preference, by taste and by priorities. There is, indeed, nothing simple about economics.

With networked, digital energy, the values we see in an offering may range from the purely practical ('I can use one simple remote control for everything from the lighting via heating to selecting the album I'm listening to while lying in the bath'), to the emotional ('It makes me feel good to know that all my energy requirement is covered by zero emission sources'), to the ideological ('I am against big monolithic internationals controlling everything and exploiting natural resources, this allows me to bypass them'), and different people will have very different reasons why they may find switching to 'smart' energy advantageous.

The big, compelling push though, for the mass market, will be price. Photovoltaics, most specifically, will become cheaper, the more they proliferate. And the cheaper they become, the more they will proliferate. They are, as we have explained, the only energy-garnering method that is subject to Moore's Law, and they are therefore set on a virtuous cycle which propels them to become one of the most important contributors to the energy pie.

There is no other force that could make markets (and for markets read: people, like you and us) undergo such radical changes so quickly. And that is a blessing, because it means we do not have to concentrate our own energies on fighting a longwinded ideological battle, which could not, in any case, be won. We may well be the only creatures on our planet capable of reason, but we are not overly rational when it comes to it. We fight wars over religious beliefs, we have lucky numbers and believe in horoscopes. It doesn't matter what your opinion is, there will always be some people who will agree with you, and there will be some who never will. But you can place your bets now that the moment solar power from photovoltaics is cheaper than, say, nuclear, there will be a scramble right across party lines and the political spectrum to have some.

But that, of course, means there needs to be a level playing field. When governments make policy decisions, they rarely have no implications on government spending. For all manner of reasons – to keep industry sectors afloat, to help people start up businesses, to keep party donors happy – governments channel funds into, and bestow tax privileges on, people they deem, in line with their openly declared policy if they are above board, as clandestine back room deals if they're not, deserving. Which is why in some countries there is a flourishing multi-million dollar lobbying industry, while in other countries people simply pay and take backhanders and bribes.

And this is mostly what we need to be wary of. Up until recently, the development of what were considered 'alternative' energy technologies was hampered not so much by a lack of public funding, but by the economic disadvantage they suffered from the extraordinarily high amounts of public funding and investments that went into fossils and nuclear in particular.

But energy does not need subsidy. There is plenty of energy around and there are more than enough private enterprises, large and small, that want to and can tap into that energy and make it accessible. All legislators need to do is facilitate this rather than obstruct it. In an open energy market, infinite energy technologies and the digital energy network will fare well.

What, then, exactly is the outlook? We have so far resisted making predictions, and we have also explained why: we know that the moment you say 'this is going to happen' and put a timescale to it, you set yourself up for being proved wrong due to all manner of unforeseen factors.

Still, with this caveat, that it may well prove wrong, our current perception is that we are looking at roughly one generation, between 25 and 30 years, to move from where we are today to where we have set out to go: always-on energy, networked, steady and clean.

We think that 'e-Day', the day when electricity from infinite sources, specifically solar, becomes cheaper than from finite sources, specifically fossils and nuclear, is not far off. Otherwise known as 'grid parity', we expect to see this in the next 4-5 years. From then on, we anticipate a dramatic acceleration in the spread of solar installations. This is likely to cause manufacturing bottlenecks and price hikes which will temporarily slow things down and have a lot of people say 'told you so!' But this 'crisis', too, will be short-lived and after another five to seven years, we expect to be back on track.

We have no way of knowing with any certainty that any of this will unfold as we suggest, but that's our educated guess. And we don't think any of this is going to happen without glitches. No radical change ever does. But we think it can happen without catastrophe, without war over resources or major disaster. And together, we believe, we can make it so.

I   AND WHAT ABOUT ME?

So what can you actually do? What is *my* role in this, you may, and we hope will, be asking.

Absolutely the most important thing is to grasp and then explain to people what this is about and what it isn't: it isn't about 'back to basics' and it's not about 'cutting back'. It's not about solar cookers (fun though are) and it's not about energy autonomy, like becoming self-sufficient on the alp or with a windmill on your roof. These are nice ideas, some of them have real charm, others are eminently useful in specific settings. Some of them we've cited ourselves as interim steps and as part of a bigger picture. There is nothing wrong with them. But they are just that: part of a much bigger picture. They are, in effect, tinkering at the fringes.

What this is about is the fundamental core body of what energy is and how we handle it. It's about networking and implementing a new energy reality. It's about bringing energy out of the industrial age and mindset into the digital age and mindset. It's categorical and it's radical. It's not about saving energy or making the share of energy from 'renewables' a bit bigger, and it's not about storage either. It's about setting off from the plane of stuffs and resources, and tapping into *inexhaustible, continuously available* sources of energy that never have and never will run out and that are so prevalent that we will have energy *abundance.* We do this by sorting energy logistics – which includes some short term storage – and thinking in terms of energy *applications.*

We have expressed in one or two places, and we are happy to re-iterate: we do not offer any solutions off the shelf. There are no *recipes.* And why not? Because we think that recipes are not the solution to any problem. They are part of the problem. They stop us from thinking for ourselves and they make things seem simple when in fact they're complicated. By making them seem simple when in fact they're complicated, they set us up for failure, because as we apply the rules or the formulas we think we can trust, we realise that in a world that has a panoply of many-layered, multifaceted problems, 'simple' solutions do not exist. Multiple layers ask for differentiation in approach. Facets require tailoring. What we need are not silver bullets that purport to blow the lock off the door into a perfect world, what we need is the imagination to invent a less imperfect world. Our appeal to you, then, goes along these lines:

- Relinquish 'scarcity'. Think, instead, in terms of 'abundance', and imagine the possibilities this opens up. If they say to you 'save energy', ask 'why?' – Is it energy we need to save, or is it invention we need to propagate in order to make use of the plentiful energy there is: think in terms of *applications.*
- Don't trust statistics. Question them, examine them. Even (especially!) the figures and graphs in this book: don't believe them. Go online and find out if there aren't other, newer, better sets of figures. Maybe figures that contradict them. There probably are. Any set of data can be used to make up any set of 'facts'. Be optimistic, but critical.
- Enjoy the theatrics, but don't take them at face value. An impressive stunt on stage or powerful editing in a film with a moving soundtrack: they make for good entertainment, but they don't in themselves make for good science. Distinguish between the story and the storytelling.
- On the subject of storytelling: ask for *good* stories. Ask for serious storytelling. It doesn't have to be earnest, there's no reason to lose your sense of humour, but take the world seriously and ask for people who tell you about it to do so too. Scaremongering and cynicism are easy. Poetry and inspiration, that's what's challenging.
- Eschew ideology. You need ethics and you need a sense of what you're about. What you don't need is people telling you how to think. What you don't need is people adhering to beliefs, purely on the grounds that that's what they do. Make your own policy and strategy. Find your own beliefs, and then test them, query them, question them.
- Be yourself and be connected. There is a 'self' in a networked world. There is autonomy and there is solidarity. You'll want and need both, because there are a lot of people in the world, and the number is growing. Embrace them. Don't be afraid of people: there is more than enough room in the world for everyone.
- Don't accept injustice. There is enough water, enough food, enough energy too. There is more than enough money. Money is a man-made thing. It can go where it's needed any time. We have civilisation, we have education, we have medicine, we have science. There is no excuse for poverty, misery and exploitation.
- Don't panic, relax. 'Fear', proverbially, 'is a bad advisor', and you don't have to be a behavioural scientist to know that that much is certainly true. The moment when you have run out of options is when you are dead. Before then, there are any number of things you can do, one of which is to keep a calm head about you and use your imagination.
- Imagine. Full stop. Don't let yourself be caught by oncoming disaster, like a rabbit in the headlights of a speeding car. You are a human being: you are capable of genius. Use it and invent the world you want to live in.

CONCLUSION

Much of our situation today is reminiscent of the problem Londoners faced at the end of the 19th century, when, based on the number of cabs, carts, coaches and buses in operation, all drawn by tens of thousands of horses, and in light of the rapid growth of the city, it was estimated that by 1950 at the latest London's streets would be nine feet deep in horse manure. The prognosis was bad. And like many of the more panic-stricken forecasts of today, it stemmed from a structurally conservative extrapolation of retro-applied statistical observations: it assumed that the future would take shape in line with the technology and mindset of the past. And therein lies a grave, fundamental error. It's impossible to predict the future. The only thing we can do is create it. Fear is a bad basis for the creative transformation of a perspective, and it's equally bad for conducting a search for new perspectives in an open-minded future.

In that sense, our CO_2 problem looks at least somewhat similar to that of the horse manure. There is indeed a problem, but the more time passes, the less it seems that the problem is due to any genuine factual reasons, but rather to a causal loop: we realise there's a problem and we try to address it with the same mindset that has got us into the problem in the first place, and before we've finished thinking our thought we realise that the problem has just got worse and we're no nearer to solving it, so we become ever more scared and ever more timid, and we begin to look backwards and to put on the brakes when we should be boldly moving forward, maybe to places where no-one has dared go before.

Being timid within an old reference system and saving resources which are going to run out one way or another is obviously not a sensible strategy for solving problems. What we need is imagination, a change of perspectives, fresh mindsets and an active, emphatic undertaking to re-articulate our world and the symbols we use to describe and handle it. And that means developing evolved reference systems with new abstractions, even new levels of abstraction. We've done it with information and data, we can do it with energy.

In the case of London manure, the internal combustion engine came along to solve the problem once and for all, and at the same time it provided people with hitherto undreamt-of levels of mobility. Of course, it also brought about problems on a grand new scale. Like we said at the very outset: we are not promising the advent of Utopia. Nevertheless, chances are good that we will be able to solve the energy and the CO_2 problem by way of technology; specifically, by way of a symbolist technology, that is a technology which creates an abstraction of energy: what is often called 'digital'. In the process, we'll be opening up unimagined potentialities on a new kind of scale. And the sooner we do so, the better, if the climatologists with their pressing and compelling warnings are anything to go by. Just how soon is really up to us. We can, all of us, do something towards creating a new energy model, whichever level we're at: we have the genius, we have the power.

APPENDIX

THE AUTHORS

LUDGER HOVESTADT is Professor for Computer Aided Architectural Design (CAAD) at the Swiss Federal Institute of Technology *(Eidgenössische Technische Hochschule, ETH)* in Zürich, Switzerland. His approach, broadly speaking, is to look for a new relationship between architecture and information technology and aims at developing a global perspective that relates to and integrates with developments in different fields such as politics and demographics, as well as technology, in a post-industrial era. He is the inventor of the digitalSTROM chip and founder of several spin-off companies in the fields of Smart Building Technology and Digital Design and Fabrication. A showcase of his recent work can be found in *Beyond the Grid – Architecture and Information Technology. Applications of a Digital Architectonic* (Birkhäuser, Basel / Boston 2009).

VERA BÜHLMANN is Professor for Architecture Theory and Philosophy of Technics at the Institute for Architectural Sciences, Technical University Vienna (Technische Universität Wien), Austria. Until September 2016, she was founding director of the Laboratory for Applied Virtuality at CAAD ETH Zürich. She has a Dr. phil. from Basel University, Switzerland, with a comparative study on contemporary media and communication theories that was published in 2014 as *Die Nachricht, ein Medium. Generische Medialität, städtische Architektonik (The Medium, a Message. Generic Mediality, Civil Architectonics)* (ambra, 2014).
    She is co-editor of the *Applied Virtuality Book Series* together with Ludger Hovestadt – among the titles are *A Quantum City* (2015), *Sheaves – When Things Are Whatever Can Be the Case* (2014), *The Metalithikum Book Series Vol. I-IV* (2012-2015). Among her other publications are many articles in journals and books, as well as, co-edited with Martin Wiedmer, *Pre-specifics. Some Comparatistic Investigations on Research in Art and Design* (JRP|Ringier, Zürich 2008).

SEBASTIAN MICHAEL thinks, writes and creates across disciplines in theatre, film, video, print and across media with a deepening interest in humans, the multiverse and quantum philosophy. He has written several stage plays, among them *The Power of Love* (Southwark Playhouse London, shortlisted for Verity Bargate Award), *Elder Latimer is in Love* (Arcola Theatre, London) and *Top Story* (The Old Vic Tunnels, London). He has written a novel, *Angel* (Optimist, 2009), as well as two short films and the feature film *The Hour of Living,* all of which he also directed.

Between 1998 and 2008, Sebastian worked extensively in a corporate set-ting, writing and devising concepts, scripts, keynotes, immersive brand experiences and experiential training programmes for companies and organisations including Vodafone, Barclays, Hewlett-Packard, Motorola, Mercedes-Benz, London Transport and the Royal Navy. Since 2009, his freelance work has shifted towards science and theory, mostly in collabo-rations with Professors Hovestadt and Bühlmann, which, apart from *A Genius Planet* also include *A Quantum City* (Birkhäuser, June 2012) and currently *An Atlas of Digital Architecture* (expected 2017/18). His online 'concept narrative in the here & now about the where, the wherefore and forever', *EDEN by FREI,* is taking shape at www.EDENbyFREI.net. Sebastian lives in London and works wherever his projects take him. @optimistlondon

FURTHER READING

We have tried, with this book, to tread a path of accessibility, and so we've focused mainly on the technical and practical aspects of what we are pro-posing. There is, however, a fairly complex theoretical framework to this, which we've touched on, and if you're interested, you will find a lot more in-depth thought and discussion at the links in the reference section below.

Here are some titles we found interesting and/or useful, espe-cially during the early stages of researching our book:

Sustainable Energy – Without the Hot Air
David JC MacKay
UIT Cambridge, 2009

Perfect Power
Robert Glavin and Kurt Yeager with Jay Stuller
McGraw-Hill, 2009

Cool It – The Skeptical Environmentalist's Guide to Global Warming
Bjørn Lomborg
Marshall Cavendish Editions, 2010 (First Published 2007)

The Third Industrial Revolution – How Lateral Power is Transforming Energy, The Economy and The World
Jeremy Rifkin
Palgrave MacMillan, 2011

Catching Fire - How Cooking Made Us Human
Richard Wrangham
Profile Books, 2009
World Changing – A User's Guide for the 21st Century

Elex Steffen, Ed
Harry N Abrams Inc, undated, probably 2009/2010
Hot, Flat and Crowded
Thomas L Friedman
Farrar, Strauss and Giroux, 2008

Ten Technologies to Save the Planet
Chris Goodall
Green Profile, 2008

Energise – A Future for Energy Innovation
James Woudhuysen & Joe Kaplinsky
Beautiful Books Limited, 2009

The Big Earth Book – Ideas and Solutions for a Planet in Crisis
James Bruges
Fragile Earth, 2007

Limits to Growth, The 30 Year Update
Donella Meadows, Jørgen Randers, Dennis Meadows
Earthscan, 2005

REFERENCE

(All web links retrieved or verified October 2016)

[I] Ludger Hovestadt is Professor for Computer Aided Architectural Design (CAAD) at ETH *(Eidgenössische Technische Hochschule)* Zürich – the Swiss Federal Institute of Technology. Vera Bühlmann was until 2016 head of the Institute of Applied Virtuality at Ludger's chair at ETH and has since taken up a professorship at TU *(Technische Universität)* Wien – the Technical University of Vienna. Sebastian Michael has been working with Ludger and Vera since 2009 on this and other projects. For a more comprehensive introduction to the authors, see Who We Are and the author biographies in the Appendix.

[II] Ivar Giaever (born 5th April 1929), Norwegian-American physicist and Nobel Prize laureate in Physics 1973: https://en.wikipedia.org/wiki/Ivar_Giaever

[III] *Ivar Giaever: Global Warming Revisited (2015)*, Lecture given as part of the Lindau Nobel Laureate Meetings series (video): http://www.mediatheque.lindau-nobel.org/videos/34729/ivar-giaever-global-warming-revisited/laureate-giaever

[IV] Homer (dates unknown, possibly some time between 1102 and 850 BCE), Greek author of the first known Western literature: https://en.wikipedia.org/wiki/Homer

[V] Hesiod (dates unknown, possibly some time between 750 and 650 BCE, or a contemporary of Homer's), Greek poet: https://en.wikipedia.org/wiki/Hesiod

[VI] Xenophanes (c. 570 – c. 475 BCE), Greek philosopher, poet and theologian, who criticised the ancient Greek poets, among them Homer and Hesiod, for anthropomorphising the gods: https://en.wikipedia.org/wiki/Xenophanes

[VII] One article that argues thus, just as an example, is: *Climate Change Will Not Be Dangerous for a Long Time,* Matt Ridley, 27 November 2015, Scientific American: https://www.scientificamerican.com/article/climate-change-will-not-be-dangerous-for-a-long-time/

[VIII] Again, simply as an example: *Climate Change Isn't World's Biggest Problem,* Alex B Berezow, 8 July 2013, Real Clear Science: http://www.realclearscience.com/articles/2013/07/08/climate_change_isnt_worlds_biggest_problem_106585.html

[IX] Gordon E. Moore (born 3 January 1929), American businessman and co-founder of the Intel Corporation: https://en.wikipedia.org/wiki/Gordon_Moore – More on Moore's Law later in this book and on Wikipedia: http://en.wikipedia.org/wiki/Moore's_law

[X] United Nations global population growth forecast to 2050: http://www.un.org/en/development/desa/news/population/2015-report.html

[XI] Sources: World History Site: http://www.worldhistorysite.com/population.html, Wikipedia: https://en.wikipedia.org/wiki/World_population_estimates and United States Census Bureau: https://www.census.gov/population/international/data/worldpop/table_history.php

[XII] This figure varies; source for this is QI, the BBC TV Quiz Show, Series I, 9.2, documented here: http://www.comedy.co.uk/guide/tv/qi/episodes/9/2/

[XIII] Charles Augustus Lindbergh (4 February 1902 – 26 August 1974), American aviator and first pilot to fly solo from New York to Paris: http://en.wikipedia.org/wiki/Charles_Lindbergh

[XIV] Albert Einstein (14 March 1879 – 18 April 1955), German-born theoretical physicist: https://en.wikipedia.org/wiki/Albert_Einstein

[XV] Albert Einstein (n.d.), BrainyQuote: https://www.brainyquote.com/quotes/quotes/a/alberteins385842.html

[XVI] Robert William 'Bob' Galvin (9 October 1922 – 11 October 2011), American businessman and from 1959 to 1986 CEO of Motorola Inc, which had been co-founded by his father in 1928: https://en.wikipedia.org/wiki/Bob_Galvin – In 2005, Galvin created the Galvin Electricity Initiative as a non-profit organisation: http://www.galvinpower.org/

[XVII] Robert Galvin and Kurt Yeager with Jay Stuller: *Perfect Power – How the Microgrid Revolution Will Unleash Cleaner, Greener and More Abundant Energy,* McGraw Hill, 2008. ISBN 978-0071548823

[XVIII] *The Fourth Revolution* (Documentary, 83', Germany, 2010) directed by Carl-A. Fechner. Film's website (English version): http://www.energyautonomy.org/index.php?article_id=21&clang=1

[XIX] Jeremy Rifkin: *The Third Industrial Revolution – How Lateral Power Is Transforming Energy, the Economy and the World,* Palgrave MacMillan, 2011. ISBN 978-0230341975

[XX] More on the Fukushima Daiichi nuclear disaster on Wikipedia: http://en.wikipedia.org/wiki/Fukushima_Daiichi_nuclear_disaster

[XXI] More on Prof Ludger Hovestadt at http://www.caad.arch.ethz.ch/blog/ludger-hovestadt/

[XXII] More on Prof Vera Bühlmann at http://tuwien.academia.edu/VeraB%C3%BChlmann

XXIII More on the ETH Future Cities Laboratory at http://www.fcl.ethz.ch/

XXIV More on Sebastian Michael at http://sebastianmichael.com

XXV For a definition and discussion of the term 'ubiquitous computing', also referred to as 'pervasive computing', see, for example: https://www.techopedia.com/definition/22702/ubiquitous-computing

XXVI There are many comparisons between the computing power of contemporary devices and 1960s space missions. For example this article from ZME Science: http://www.zmescience.com/research/technology/smartphone-power-compared-to-apollo-432/

XXVII The EIA (United States Energy Information Administration) *International Energy Outlook 2016* Report provides comprehensive data and forecasts on energy consumption through 2040: https://www.eia.gov/forecasts/ieo/

XXVIII *An Inconvenient Truth* (Documentary, 96', USA, 2006) directed by Davis Guggenheim. At IMDb: http://www.imdb.com/title/tt0497116/

XXIX Al Gore: *An Inconvenient Truth – The Planetary Emergency of Global Warming and What We Can Do About It,* Rodale, 2006. ISBN 1-59486-567-1 (Some editions may not feature the interactive pull-out element mentioned in the text.)

XXX A brief background and history to Aizo and digitalSTROM is on their website: http://www.digitalstrom.com/en/Company/digitalSTROM-AG/Briefly/

XXXI Tom Lehrer (born 9 April 1928), American singer-songwriter, satirist and mathematician: http://en.wikipedia.org/wiki/Tom_Lehrer and his own website: http://www.tomlehrer.org/

XXXII Wolfgang Amadeus Mozart (27 January 1756 – 5 December 1791), Austrian composer: http://en.wikipedia.org/wiki/Wolfgang_Amadeus_Mozart

XXXIII *List of Countries by Life Expectancy* compiled by the World Health Organisation (WHO) for 2015, published on Wikipedia: https://en.wikipedia.org/wiki/List_of_countries_by_life_expectancy

XXXIV Source: CIA World Factbook, compiled and published by Index Mundi: http://www.index-mundi.com/g/r.aspx?t=100&v=39&l=en

XXXV Oscar Wilde (16 October 1854 – 30 November 1900), Irish playwright, novelist, essayist and poet: https://en.wikipedia.org/wiki/Oscar_Wilde

XXXVI As of June 2016, there were 2.2 million apps available for the Android operating system, while iPhone users had a choice of 2 million apps: https://www.statista.com/statistics/276623/number-of-apps-available-in-leading-app-stores/

XXXVII John Steinbeck (27 February 1902 – 20 December 1968), American author of fiction and non-fiction and Nobel Prize laureate for Literature 1962.

XXXVIII John Steinbeck (n.d.), BrainyQuote: https://www.brainyquote.com/quotes/quotes/j/johnsteinb121626.html

XXXIX More on the Solar Constant at Encyclopaedia Britannica: http://www.britannica.com/EBchecked/topic/552889/solar-constant or on Wikipedia: http://en.wikipedia.org/wiki/Solar_constant

XL Source: *Key World Energy Statistics 2016,* International Energy Agency (IEA): https://www.iea.org/publications/freepublications/publication/key-world-energy-statistics.html

XLI Source: Wikipedia: http://en.wikipedia.org/wiki/File:Solar_land_area.png

XLII Source: US Energy Information Administration (EIA): https://www.eia.gov/todayinenergy/detail.php?id=26212

XLIII Source: *Key World Energy Statistics 2016,* International Energy Agency (IEA): https://www.iea.org/publications/freepublications/publication/key-world-energy-statistics.html

XLIV *List of Countries by Road Network Size* compiled on Wikipedia: https://en.wikipedia.org/wiki/List_of_countries_by_road_network_size

XLV *Agricultural Land Data* compiled on Wikipedia: https://en.wikipedia.org/wiki/Agricultural_land

XLVI Source: Scientific American: http://blogs.scientificamerican.com/guest-blog/2011/03/16/smaller-cheaper-faster-does-moores-law-apply-to-solar-cells/

XLVII The philosophical concept of rhizomes was developed by Gilles Deleuze (18 January 1925 – 4 November 1995) and Félix Guattari (30 April 1930 – 29 August 1992). A brief introduction on Wikipedia: https://en.wikipedia.org/wiki/Rhizome_(philosophy)

XLVIII The Great Pacific Garbage Patch has been known about for several decades, but its dimensions are still good for surprises: *'Great Pacific Garbage Patch' Far Bigger Than Imagined, Aerial Survey Shows,* Oliver Milman, 4 October 2016, The Guardian: https://www.theguardian.com/environment/2016/oct/04/great-pacific-garbage-patch-ocean-plastic-trash. And the Wikipedia page on the patch: https://en.wikipedia.org/wiki/Great_Pacific_garbage_patch

XLIX Source BBC News: http://www.bbc.co.uk/news/world-asia-pacific-13289607

L Thomas Edison (11th February 1847 - 18th October 1931), American inventor and businessman: http://en.wikipedia.org/wiki/Thomas_Edison and a dedicated website: http://www.thomasedison.com/

LI *The King's Speech* (Feature Film, 118', UK, 2010) directed by Tom Hooper

LII The figure of 2 billion viewers is difficult to accurately ascertain and remains an estimate. See for example *Royal Wedding Facts and Figures,* 29 April 2011, Daily Telegraph: http://www.telegraph.

co.uk/news/uknews/royal-wedding/8483199/Royal-wedding-facts-and-figures.html and, for more details and a discussion of the viewing figure estimates: Wikipedia https://en.wikipedia.org/wiki/Wedding_of_Prince_William_and_Catherine_Middleton

LIII European Organization for Nuclear Research (CERN): http://public.web.cern.ch/public/

LIV *Higgs-Boson Like Particle Discovery Claimed at LHC,* Paul Rincon, 4 July 2012, BBC Science & Environment: http://www.bbc.co.uk/news/world-18702455

LV Malala Yousafzai (born 12 July 1997), Pakistani education activist and Nobel Peace Prize laureate 2014: https://en.wikipedia.org/wiki/Malala_Yousafzai

LVI *"Après nous le déluge"* – attributed to Madame de Pompadour (29 December 1721 – 15 April 1764), mistress to King Louis XV of France http://en.wikipedia.org/wiki/Madame_de_Pompadour

LVII *Urban Population (% of Total), United Nations, World Urbanization Prospects* collated by The World Bank: http://data.worldbank.org/indicator/SP.URB.TOTL.IN.ZS

LVIII More on the city and definitions of a city around the world: *What Is a City? What Is Urbanization?,* Carl Haub, October 2009, The Population Reference Bureau: http://www.prb.org/Publications/Articles/2009/urbanization.aspx

LIX Source: *World Population Prospects – 2015 Revision* United Nations, 2015: https://esa.un.org/unpd/wpp/publications/files/key_findings_wpp_2015.pdf

LX A compact table showing world population growth since the year 1 CE: http://geography.about.com/od/obtainpopulationdata/a/worldpopulation.htm

LXI This is a rough estimate we arrived at ourselves when we started writing this book in 2009. It's an educated, but most likely conservative guess as the number is certain to have increased since then.

LXII The 18 Euro cents is a European average. Prices vary greatly across the EU, ranging from about 10 cents in Bulgaria to up to 30 cents in Denmark and Germany. Source: *Electricity Prices in Europe,* Evan Lamos, 18 December 2015 (updated 20 February 2016) on EurActiv: https://www.euractiv.com/section/energy/video/electricity-prices-in-europe/

LXIII George Westinghouse (6 October 1846 – 12 March 1914), American entrepreneur and engineer: https://en.wikipedia.org/wiki/George_Westinghouse

LXIV Michael Faraday (22 September 1791 – 25 August 1867), English scientist: http://en.wikipedia.org/wiki/Michael_Faraday and on BBC History: http://www.bbc.co.uk/history/historic_figures/faraday_michael.shtml

LXV Joseph Henry (17 December 1797 – 13 May 1878), American scientist: http://en.wikipedia.org/wiki/Joseph_Henry

LXVI Alessandro Volta (18 February 1745 – 5 March 1827), Italian physicist and chemist: http://en.wikipedia.org/wiki/Alessandro_Volta

LXVII André-Marie Ampère (20 January 1775 – 10 June 1836), French physicist and mathematician: http://en.wikipedia.org/wiki/Andr%C3%A9-Marie_Amp%C3%A8re

LXVIII Source: *Breakdown of Electricity Generation by Energy Source,* The Shift Project Data Portal: http://www.tsp-data-portal.org/Breakdown-of-Electricity-Generation-by-Energy-Source#tspQvChart (Some sources give marginally varying percentages, owing to calculation methods and rounding differences.)

LXIX *List of Largest Hydroelectric Power Stations,* Wikipedia: https://en.wikipedia.org/wiki/List_of_largest_hydroelectric_power_stations

LXX *WWEA Half-Year Report: World Wind Capacity Reached 456 GW,* October 2016, World Wind Energy Association: http://www.wwindea.org/wwea-half-year-report-worldwind-wind-capacity-reached-456-gw/

LXXI Source: *Electric Power Consumption per Capita (kWh),* The World Bank: http://data.worldbank.org/indicator/EG.USE.ELEC.KH.PC

LXXII Jawed Karim (born 28 October 1978), German-American internet entrepreneur and co-founder of YouTube: https://en.wikipedia.org/wiki/Jawed_Karim

LXXIII *Me at the Zoo,* the first ever video to be uploaded to YouTube: https://www.youtube.com/watch?v=jNQXAC9IVRw

LXXIV Yakov Lapitsky is now a Professor at the University of Toledo: http://www.che.utoledo.edu/lapitsky.htm

LXXV Chad Hurley (born 24 January 1977), American entrepreneur and co-founder of YouTube: http://www.che.utoledo.edu/lapitsky.htm

LXXVI Steve Chen (born 18th August 1978), Taiwanese-born American entrepreneur and co-founder of YouTube: https://en.wikipedia.org/wiki/Steve_Chen

LXXVII A brief background to this story on Wikipedia: https://en.wikipedia.org/wiki/It%27s_The_Sun_Wot_Won_It

LXXVIII More on the *News International Phone Hacking Scandal* at Wikipedia: https://en.wikipedia.org/wiki/News_International_phone_hacking_scandal

LXXIX The Wikipedia entry on Wikipedia: http://www.wikipedia.org

LXXX Theo Angelopoulos (27 April 1935 – 24 January 2012), Greek filmmaker: http://en.wikipedia.org/wiki/Theodoros_Angelopoulos

LXXXI *NME Editor Luke Lewis Apologizes for Ed Sheeran Outburst,* Alex Langlands, 17 January 2012, MusicFeeds: http://musicfeeds.com.au/news/nme-editor-luke-lewis-apologizes-for-ed-sheeran-outburst/

LXXXII Sir Francis Bacon (22 January 1561 – 9 April 1626), English philosopher, scientist, author and statesman: https://en.wikipedia.org/wiki/Francis_Bacon

LXXXIII Gottfried Wilhelm Leibniz (1 July 1646 [O.S. 21 June] – November 14, 1716), German philosopher and polymath: https://en.wikipedia.org/wiki/Gottfried_Wilhelm_Leibniz

LXXXIV More on binary numbers at Wikipedia: https://en.wikipedia.org/wiki/Binary_number

LXXXV For more on quantum computing, including a short video: *Massive Disruption Is Coming With Quantum Computing,* Peter Diamandis, 10 October 2016, Singularity Hub: http://singularityhub.com/2016/10/10/massive-disruption-quantum-computing/

LXXXVI The Norwegian Pearl on Wikipedia: https://en.wikipedia.org/wiki/Norwegian_Pearl

LXXXVII One of the news items covering the story at the time: *German Power Firm to Risk Another Switch-off,* cro/Reuters/dpa, 6 November 2006, Der Spiegel Online: http://www.spiegel.de/international/european-blackout-german-power-firm-to-risk-another-switch-off-a-446770.html

LXXXVIII More on Wilfried Beck at http://beck24.com/index.html

LXXXIX Bill Gates (born 28 October 1955), American business magnate and co-founder of Microsoft: https://en.wikipedia.org/wiki/Bill_Gates

XC Alan Turing (23 June 1912 – 7 June 1954), English computer scientist, mathematician and cryptanalyst, and pioneer in computer science and artificial intelligence: https://en.wikipedia.org/wiki/Alan_Turing

XCI More on the Universal Turing Machine at Wikipedia: http://en.wikipedia.org/wiki/Universal_Turing_machine

XCII The digitalSTROM Alliance has since been subsumed into the digitalSTROM company and brand, but still exists as a 'technology community' of "organizations from industry, research and application to integrate with digitalSTROM": https://www.digitalstrom.org/en/

XCIII Source: Wikipedia https://en.wikipedia.org/wiki/App_Store_(iOS) – see also note 33 above.

XCIV Neil Papworth (born 1969), British software architect, designer and developer who sent the first ever SMS text message: https://en.wikipedia.org/wiki/Neil_Papworth

XCV *OMG! The First Text Message Sent 23 Years Ago Today,* Ann Brenoff, 3 December 2015, The Huffington Post: http://www.huffingtonpost.com/entry/omg-happy-23rd-birthday-to-the-text-message_us_565dcdefe4b079b2818bcb5a

XCVI *Power Surge Points to Huge Royal Wedding Ratings,* John Plunkett, 29 April 2011, The Guardian: https://www.theguardian.com/media/2011/apr/29/power-surge-royal-wedding-ratings

XCVII Source: GridWatch: http://gridwatch.co.uk/

XCVIII *Olympics Sap China's Power Supply,* Michael Lelyveld, 21 July 2008, Radio Free Asia (RFA): http://www.rfa.org/english/commentaries/energy_watch/China-energy-07212008133940.html

XCIX *Where Does the Electricity and the Gas Used in Belgium Come From?* Energuide, 2016: http://www.energuide.be/en/questions-answers/where-does-the-electricity-and-the-gas-used-in-belgium-come-from/4/

C *Belgium's Highways Shine Into Space – But for How Long?* Phillip Saure, 13 July 2011, Times of Malta: http://www.timesofmalta.com/articles/view/20110713/world/Belgium-s-highways-shine-into-space-but-for-how-long-.375311

CI Source: The World Factbook, United States Central Intelligence Agency (CIA): https://www.cia.gov/library/publications/the-world-factbook/rankorder/2236rank.html and https://www.cia.gov/library/publications/the-world-factbook/rankorder/2233rank.html

CII *Negative Wholesale Power Prices: Why They Occur and What to Do About Them,* Maria Woodman, 2011, Economics Department, New York University: http://www.usaee.org/usaee2011/submissions/OnlineProceedings/Online%20Proceeding%20Paper%20-%20Maria%20Woodman.pdf

CIII Source: *Electricity Production and Spot Prices in Germany 2014,* Johannes Mayer, 31 December 2014, Fraunhofer Institute for Solar Energy Systems ISE: https://www.ise.fraunhofer.de/de/downloads/pdf-files/data-nivc-/folien-electricity-spot-prices-and-production-data-in-germany-2014-engl.pdf

CIV Source: Smart Energy GB: https://www.smartenergygb.org/en/the-bigger-picture/about-the-rollout

CV Courtesy of digitalSTROM http://www.digitalstrom.com/System/UEbersicht/

CVI Source: The World Factbook, United States Central Intelligence Agency (CIA): https://www.cia.gov/library/publications/the-world-factbook/geos/sz.html

CVII Energy statistics source: *Schweizerische Gesamtenergiestatistik 2015,* Bundesamt für Energie, Schweizerische Eidgenossenschaft: http://www.bfe.admin.ch/themen/00526/00541/00542/00631/index.html?lang=de&dossier_id=00763

CVIII Audi's own website on the e-gas project: http://www.audi.com/corporate/en/sustainability/we-live-responsibility/product/audi-e-gas-project.html

CIX Source: The World Factbook, United States Central Intelligence Agency (CIA): https://www.cia.gov/library/publications/the-world-factbook/geos/in.html

CX Source: *Electric Power Consumption per Capita (kWh),* The World Bank: http://data.worldbank.org/indicator/EG.USE.ELEC.KH.PC

CXI *India's Per Capita Electricity Consumption Touches 1010 kWh,* Utpal Bhaskar, 23 October 2016, Live Mint: http://www.livemint.com/Industry/jqvJpYRpSNyldcuUlZrqQM/Indias-per-capita-electricity-consumption-touches-1010-kWh.html

CXII Source: Press Information Bureau, Government of India: http://pib.nic.in/newsite/PrintRelease.aspx?relid=103216 (These figures are for the period 2011-2012, but there is some consistency in this divergent pattern, which will not have changed significantly over the subsequent period.)

CXIII Source: The World Factbook, United States Central Intelligence Agency (CIA): https://www.cia.gov/library/publications/the-world-factbook/geos/in.html

CXIV *Median Household Size Drops Below 4 in Cities,* Rukmini Shrinivasan, 25 March 2012, The Times of India: http://timesofindia.indiatimes.com/india/Median-household-size-drops-below-4-in-cities/articleshow/12397117.cms

CXV *In Rural India, Solar-Powered Microgrids Show Mixed Success,* Fred Pearce, 14 January 2016, Yale Environment 360: http://e360.yale.edu/feature/in_rural_india_solar-powered_microgrids_show_mixed_success/2948/

CXVI The Tata Power Solar website on solar microgrids: http://www.tatapowersolar.com/Solar-Microgrid

CXVII Source: The World Factbook, United States Central Intelligence Agency (CIA): https://www.cia.gov/library/publications/the-world-factbook/geos/et.html

CXVIII Source: *Extension of Deforestation in Ethiopia: A Review,* Dr S Srinivasan, February 2014, EPRA International Journal of Economic and Business Review: http://eprawisdom.com/jpanel/upload/articles/1022amdrsrinivasan.pdf

CXIX *Awareness of Health Effects of Cooking Smoke Among Women in the Gondar Region of Ethiopia: a Pilot Survey,* M Edelstein, E Pitchforth, G Asres, M Silverman, N Kulkami, 18 July 2008, US National Center for Biotechnology Information (NCBI): https://www.ncbi.nlm.nih.gov/pmc/articles/PMC2491593/

CXX Source: The World Factbook, United States Central Intelligence Agency (CIA): https://www.cia.gov/library/publications/the-world-factbook/geos/et.html

CXXI Source: NationMaster: http://www.nationmaster.com/country-info/profiles/Ethiopia/Transport

CXXII *Power Generation Begins at 1,870-MW Gibe III Hydroelectric Project in Ethiopia,* Gregory B Poindexter, 14 October 2015, HydroWorld: http://www.hydroworld.com/articles/2015/10/power-generation-begins-at-1-870-mw-gibe-iii-hydroelectric-project-in-ethiopia.html

CXXIII More on the Gilgel Gibe III Dam and the Gibe Cascade on Wikipedia: http://en.wikipedia.org/wiki/Gilgel_Gibe_III_Dam

CXXIV David Hockney (born 9 July 1937), British artist: https://en.wikipedia.org/wiki/David_Hockney

CXXV *David Hockney: A Bigger Picture* ran at the Royal Academy of Arts from 21 January to 9 April 2012: https://www.royalacademy.org.uk/exhibition/david-hockney-a-bigger-picture

CXXVI *David Hockney's iPad Art,* Martin Gayford, 20 October 2010, The Daily Telegraph: http://www.telegraph.co.uk/culture/art/art-features/8066839/David-Hockneys-iPad-art.html

CXXVII *Bar Code Founder Honored for His Vision 50 Years Ago,* Sean Flynn, 14 June 2011, Newport Daily News: http://sailing.mit.edu/wikiupload/b/bb/50th_Anniversary_Article.pdf

CXXVIII More on KarTrak at Wikipedia: https://en.wikipedia.org/wiki/KarTrak

CXXIX More on yoghurt at Wikipedia: https://en.wikipedia.org/wiki/Yogurt

CXXX One of the most comprehensive and interesting sources for global wealth and health statistics is Hans Rosling's Gapminder at https://www.gapminder.org/

CXXXI The Wikipedia article on this fascinating phenomenon: http://en.wikipedia.org/wiki/Wave%E2%80%93particle_duality

CXXXII Michael Grätzel (born 11 May 1944), Swiss inventor and photovoltaic solar technology pioneer: https://en.wikipedia.org/wiki/Michael_Gr%C3%A4tzel

CXXXIII *Swiss Solar Innovator Wins Millennium Technology Prize,* 9 June 2010, BBC News: http://www.bbc.co.uk/news/10276652

CXXXIV Michael Grätzel: *Dye-sensitized Solar Cells,* July 2003, Journal of Photochemistry and Photobiology, on ResearchGate: https://www.researchgate.net/publication/232413380_Dye-Sensitized_Solar_Cells and the Wikipedia entry on DSSCs: https://en.wikipedia.org/wiki/Dye-sensitized_solar_cell

CXXXV *Organic Solar Cells With High Electric Potential for Portable Electronics,* 11 October 2012, Science Daily: http://www.sciencedaily.com/releases/2012/10/121011134738.htm

CXXXVI Source: *Commodity Profile for Rice – October 2016,* Department of Agriculture, Cooperation & Farmers Welfare, Government of India: http://agricoop.nic.in/imagedefault1/Rice_Apr15.pdf

CXXXVII Source: *Solar Photovoltaic Electricity Generation,* Strategic Energy Technologies Information System (SETIS), European Commission: https://setis.ec.europa.eu/system/files/Technology_Information_Sheet_Solar_Photovoltaic.pdf

CXXXVIII *Energy Efficient Buildings Are Vital to Sustainability,* Christian Kornevall, 1 April 2011, The Guardian: https://www.theguardian.com/sustainable-business/energy-efficient-buildings1

CXXXIX Source: US Energy Information Administration (EIA), 19 November 2015: http://www.eia.gov/todayinenergy/detail.php?id=23832

CXL *2,000 Watt Society,* Kalle Huebner, 2 February 2009, Our World, United Nations University: https://ourworld.unu.edu/en/2000-watt-society

CXLI Zaha Hadid (31 October 1950 – 31 March 2016), Iraqi-born British architect: https://en.wikipedia.org/wiki/Zaha_Hadid

CXLII Frank Gehry (born 28 February 1929), Canadian-born American architect: https://en.wikipedia.org/wiki/Frank_Gehry

CXLIII Rem Koolhaas (born 17 November 1944), Dutch architect: https://en.wikipedia.org/wiki/Rem_Koolhaas

CXLIV More on architectural styles in this compilation by the Royal Institute of British Architects (RIBA): https://www.architecture.com/Explore/ArchitecturalStyles/Architecturalstyles.aspx

CXLV Rudolf Clausius (2 January 1822 – 24 August 1888), German physicist and mathematician: https://en.wikipedia.org/wiki/Rudolf_Clausius

CXLVI The Wikipedia article on Exergy: https://en.wikipedia.org/wiki/Exergy

CXLVII Source: *A World-Leader in Wind Energy,* November 2015, Denmark – The Official Website of Denmark: http://denmark.dk/en/green-living/wind-energy/

CXLVIII A collection of stories on wind power, collated by The Guardian: https://www.theguardian.com/environment/windpower

CXLIX *Carnegie Installs Final Unit in World's First Grid-Connected Wave Energy Plant,* Sophie Vorrath, 12 March 2015, RenewEconomy: http://reneweconomy.com.au/2015/carnegie-installs-final-unit-in-worlds-first-grid-connected-wave-energy-plant-27462

CL *New Generation Wave Energy: Could it Provide one Third of Australia's Electricity?* Giles Parkinson, 1 December 2015, The Guardian: https://www.theguardian.com/sustainable-business/2015/dec/01/new-generation-wave-energy-could-it-provide-one-third-of-australias-electricity

CLI More on wave and ocean power technologies at Alternative Energy News: http://www.alternative-energy-news.info/technology/hydro/wave-power/

CLII *World's First Lagoon Power Plants Unveiled in UK,* Roger Harrabin, 2 March 2015, BBC News: http://www.bbc.co.uk/news/science-environment-31682529

CLIII *Sihwa Lake Tidal Power Plant, Gyeonggi Province, South Korea,* Sonal Patel, 12 January 2015, Power Magazine: http://www.powermag.com/sihwa-lake-tidal-power-plant-gyeonggi-province-south-korea/

CLIV *Pentland Firth Tidal Power Plant, Scotland, United Kingdom,* [nd], Power-Technology.com: http://www.power-technology.com/projects/pentland-firth-tidal-power-plant-scotland/

CLV *Tidal Energy Project in the Bay of Fundy,* Natural Resources Canada, Government of Canada: http://www.nrcan.gc.ca/energy/funding/current-funding-programs/cef/4955

CLVI Tidal Devices explained by the European Marine Energy Centre (EMEC) in Orkney: http://www.emec.org.uk/marine-energy/tidal-devices/

CLVII Source: International Energy Agency (IEA): http://www.iea.org/topics/renewables/subtopics/hydropower/

CLVIII Hydropower explained by the US Energy Information Administration: http://www.eia.gov/energyexplained/?page=hydropower_home

CLIX Source: World Energy Council: https://www.worldenergy.org/data/resources/resource/geothermal/

CLX Source: *How Geothermal Energy Works,* Union of Concerned Scientists: http://www.ucsusa.org/clean_energy/our-energy-choices/renewable-energy/how-geothermal-energy-works.html#.WAeuBpMrKL4

CLXI *Some Geothermal Electricity Production Basics* from the National Renewable Energy Laboratory (NREL), US Department of Energy: https://www.nrel.gov/workingwithus/re-geo-elec-production.html

CLXII Source: Wikipedia: https://en.wikipedia.org/wiki/Concentrated_solar_power

CLXIII *Dubai Is Building the World's Largest Concentrated Solar Power Plant,* George Dvorsky, 6 June 2016, Gizmodo: http://gizmodo.com/dubai-is-building-the-worlds-largest-concentrated-solar-1780781150

[CLXIV] Source: Wikipedia: https://en.wikipedia.org/wiki/Concentrated_solar_power

[CLXV] An explanation of different Concentrated Solar Power (CSP) technologies from Solar Cell Central: http://solarcellcentral.com/csp_page.html

[CLXVI] *Artificial Photosynthesis to Power the Future,* Tim Maverick, 10 February 2016, Wall St Daily: http://www.wallstreetdaily.com/2016/02/10/artificial-photosynthesis-energy/

[CLXVII] Hermann Emil Fischer (9 October 1852 – 15 July 1919), German chemist and Nobel Prize in Chemistry laureate 1902: https://en.wikipedia.org/wiki/Hermann_Emil_Fischer

[CLXVIII] For example: *Artificial Photosynthesis: How Renewable Fuels Will Make Oil Obsolete,* Chris Goodall, 22 August 2016, The Ecologist: http://www.theecologist.org/blogs_and_comments/commentators/2988018/artificial_photosynthesis_how_renewable_fuels_will_make_oil_obsolete.html

[CLXIX] *How Artificial Photosynthesis Works* explained by Julia Layton in How Stuff Works: http://science.howstuffworks.com/environmental/green-tech/energy-production/artificial-photosynthesis2.htm

[CLXX] Ian McEwan, *Solar,* Random House, 2010. ISBN 0-224-09049-6

[CLXXI] Forecasts by the US Energy Information Administration (EIA) for installations entering service in 2022 put the cost of 'advanced nuclear' over the life cycle of a plant at between 56% to 96% higher than for onshore wind or photovoltaic solar: http://www.eia.gov/forecasts/aeo/pdf/electricity_generation.pdf

[CLXXII] The Wikipedia article on the Fukushima Daiichi Nuclear Disaster: https://en.wikipedia.org/wiki/Fukushima_Daiichi_nuclear_disaster

[CLXXIII] Source: Wikipedia: https://en.wikipedia.org/wiki/Deaths_due_to_the_Chernobyl_disaster

[CLXXIV] This estimate and the following calculations involving the cost of photovoltaic solar panels are our own estimates.

[CLXXV] *Switzerland Decides on Nuclear Phase-Out,* James Kanter, 25 May 2011, The New York Times: http://www.nytimes.com/2011/05/26/business/global/26nuclear.html

[CLXXVI] Source: *Economics of Nuclear Power FAQs,* Nuclear Energy Agency (NEA), Organisation for Economic Co-ordination and Development (OECD): https://www.oecd-nea.org/news/press-kits/economics-FAQ.html#1

[CLXXVII] *How Long Will the World's Uranium Supplies Last?* Steve Fetter, [n.d], Scientific American: https://www.scientificamerican.com/article/how-long-will-global-uranium-deposits-last/

[CLXXVIII] You can see *Il Distendino* in action here: https://vimeo.com/972849

[CLXXIX] Jean Tinguely (22 May 1925 – 30 August 1991), Swiss artist and maker of sculptural machines: http://en.wikipedia.org/wiki/Jean_Tinguely

[CLXXX] You can see the *Fasnachtsbrunnen* in action here: https://vimeo.com/837607

[CLXXXI] *How Tesla's Batteries Can Change the Solar Power Game,* Hal Hodson, 10 August 2016, New Scientist: https://www.newscientist.com/article/mg23130860-200-how-teslas-batteries-can-change-the-solar-power-game/

[CLXXXII] Ernst Ulrich von Weizsäcker (born 25 June 1939), German scientist and politician: https://en.wikipedia.org/wiki/Ernst_Ulrich_von_Weizs%C3%A4cker Author of the book *Factor Four: Doubling Wealth, Halving Resource Use – A Report to the Club of Rome,* Earthscan Publications, 1998. ISBN 9781853834066

[CLXXXIII] Al Gore: *Our Choice: A Plan to Solve the Climate Crisis,* Rodale Books, 2009. ISBN 978-1594867347

[CLXXXIV] A brief, '5-minute' introduction to The Global Marshall Plan on its own website: http://www.globalmarshallplan.org/en/5-minutes

[CLXXXV] Source: The World Bank: http://www.worldbank.org/en/topic/poverty/overview#1

[CLXXXVI] The 50,000 people a day figure (18 million people a year) stems from the Wikipedia article on poverty – https://en.wikipedia.org/wiki/Poverty – but the sources cited for it are somewhat out of date. Since the United Nations *Development Report 2003* and the World Health Organisation (WHO) *World Health Report 1999,* the number of people living in extreme poverty has declined in both absolute and relative terms. That notwithstanding, *11 Facts About Poverty* on the DoSomething.org platform cites a UNICEF 2014 report as estimating the number of child deaths alone due to poverty at 22,000 each day, while poverty.com cites the World Food Program, Oxfam and UNICEF as its sources for an estimated figure of 21,000 deaths a day due to hunger and hunger-related causes. Although an exact figure is virtually impossible to arrive at, it is still entirely plausible that with 2.4 billion people "still lacking improved sanitation" in 2015 (UNICEF: http://www.unicef.org/publications/index_82419.html), for example, poverty is a major contributing factor to the premature deaths of less than a percent of them.

[CLXXXVII] The Wikipedia article on the Diffusion of Innovations: http://en.wikipedia.org/wiki/Diffusion_of_innovations

[CLXXXVIII] Steve Jobs (24 February 1955 – 5 October 2011), American inventor and entrepreneur and co-founder and CEO of Apple Inc: https://en.wikipedia.org/wiki/Steve_Jobs

CLXXXIX Source: Wikipedia World Population Estimates: https://en.wikipedia.org/wiki/World_population_estimates

CXC The great success of German energy farmers has a lot to do with the country's energy policies, which are not without their critics, as this story illustrates: *German Farmers Reap Benefits of Harvesting Renewable Energy,* Jeevan Vasagar, 2 December 2013, Financial Times: https://www.ft.com/content/f2bc3958-58f4-11e3-9798-00144feabdco

CXCI *Saudi Arabia Sees its Oil Reserves Lasting Another 70 Years,* Anthony Dipaola, 11 October 2016, Bloomberg: http://www.bloomberg.com/news/articles/2016-10-11/saudi-arabia-sees-its-oil-reserves-lasting-another-70-years

CXCII Donella H Meadows, Dennis L Meadows, Jørgen Randers, William W Behrens III: *The Limits to Growth,* Universe Books, 1972. ISBN 0876631650 and Donella H Meadows, Dennis L Meadows, Jørgen Randers: *Limits to Growth: The 30-Year Update,* Chelsea Green, 2004. ISBN 978-1931498586

CXCIII More on Hermes at Wikipedia: http://en.wikipedia.org/wiki/Hermes

CXCIV In 1983 the metre was redefined as the distance light travels in a vacuum in 1/ 299 792 458 seconds. More at Wikipedia: https://en.wikipedia.org/wiki/Metre

CXCV Diophantus (approximately 201-215 – 285-299 CE), Greek mathematician known as the 'father of algebra': https://en.wikipedia.org/wiki/Diophantus

CXCVI Euclid of Alexandria (fl. 300 BCE), Greek mathematician known as the 'father of geometry': https://en.wikipedia.org/wiki/Euclid

CXCVII *The History of Zero – How Was Zero Discovered,* Nils-Bertil Wallin, 2002, Yale Global Online: http://yaleglobal.yale.edu/about/zero.jsp

CXCVIII Source: Wikipedia: https://en.wikipedia.org/wiki/Pi

CXCIX The proof for this assertion is given in Cantor's Diagonal Argument, about which more on Wikipedia: https://en.wikipedia.org/wiki/Cantor%27s_diagonal_argument

CC Leonardo da Vinci (15 April 1452 – 2 May 1519), Italian polymath, artist, inventor and engineer: https://en.wikipedia.org/wiki/Leonardo_da_Vinci

CCI The eruption of the Icelandic Eyjafjallajökull volcano led to the grounding of European air traffic from the 14th to the 21st April 2010: *How Vulcano Chaos Unfolded: in Graphics,* BBC News: http://news.bbc.co.uk/1/hi/world/europe/8634944.stm and on Wikipedia: https://en.wikipedia.org/wiki/2010_eruptions_of_Eyjafjallaj%C3%B6kull

CCII Hannibal (247 – c. 182 BCE), Punic military commander, who famously led an army across the Pyrenees and over the Alps to confront the Romans in Italy: https://en.wikipedia.org/wiki/Hannibal

CCIII *Hannibal's Crossing of the Alps* on Wikipedia: https://en.wikipedia.org/wiki/Hannibal%27s_crossing_of_the_Alps

CCIV Source: Reuters http://www.reuters.com/article/2011/05/04/us-energy-ipcc-idUSTRE7432 5N20110504?feedType=RSS&feedName=everything&virtualBrandChannel=11563